THE CHANGING GOVERNANCE OF RENEWABLE NATURAL RESOURCES IN NORTHWEST RUSSIA

The Changing Governance of Renewable Natural Resources in Northwest Russia

Edited by
SOILI NYSTÉN-HAARALA
University of Joensuu, Finland

Routledge
Taylor & Francis Group

LONDON AND NEW YORK

First published 2009 by Ashgate Publishing

Published 2016 by Routledge
2 Park Square, Milton Park, Abingdon, Oxfordshire OX14 4RN
711 Third Avenue, New York, NY 10017, USA

First issued in paperback 2016

Routledge is an imprint of the Taylor & Francis Group, an informa business

British Library Cataloguing in Publication Data
The changing governance of renewable natural resources in
    northwest Russia
    1. Renewable natural resources - Law and legislation -
    Russia (Federation) 2. Renewable natural resources -
    Russia, Northern - Management 3. Capitalism - Russia,
    Northern
    I. Nysten-Haarala, Soili
    346.4'70467

Library of Congress Cataloging-in-Publication Data
The changing governance of renewable natural resources in northwest Russia / [edited] by Soili Nystin-Haarala.
        p. cm.
    Includes index.
    ISBN 978-0-7546-7531-0
    1. Natural resources--Russia, Northwestern--Management. 2. Environmental management--Russia, Northwestern. I. Nystin-Haarala, Soili.

    HC340.12.Z7N67385 2008
    333.70947'1--dc22

                                                                2008041328

    ISBN 13: 978-1-138-27846-2 (pbk)
    ISBN 13: 978-0-7546-7531-0 (hbk)

Contents

List of Figures and Tables   *vii*
List of Contributors   *ix*
Preface   *xiii*
List of Abbreviations   *xv*

1   Introduction   1
       Soili Nystén-Haarala

**PART I: NATIONAL POLICIES AND A TRANSITION
            TO A MARKET ECONOMY**

2   Institutions, Interest Groups and Governance of Natural Resources
       in Russia   9
       Soili Nystén-Haarala and Juha Kotilainen

3   The Task of Macroeconomic Policy in Generating Trust
       in Russia's Development   31
       Stefan Walter

4   Russian Forest Regulation and the Integration of Sustainable
       Forest Management   55
       Minna Pappila

5   Fishery Governance in Northwest Russia   77
       Larissa Riabova and Lyudmila Ivanova

6   The Struggle for the Ownership of Pulp and Paper Mills   105
       Anna-Maija Matilainen

**PART II: CASE STUDIES ON DIFFERENT ASPECTS
            OF GOVERNANCE**

7   Re-Territorializing the Russian North Through Hybrid
       Forest Management   131
       *Juha Kotilainen, Antonina A. Kulyasova, Ivan P. Kulyasov
       and Svetlana S. Pchelkina*

8   Construction of Trust in Russian Mill Towns   149
        Jarmo Kortelainen and Soili Nystén-Haarala

9   Conflict as a Form of Governance: The Market Campaign
        to Save the Karelian Forests   169
        Maria Tysiachniouk

10   Transformation of Nature Management in Pomorie:
        Fishing Villages on the Onega Peninsula of the White Sea   197
        Antonina A. Kulyasova and Ivan P. Kulyasov

PART III: THE INTERNATIONAL AND GLOBAL IMPACT
ON NATIONAL ENVIRONMENTAL POLICY
AND LOCAL FORESTRY AND FISHERY

11   Local Adaptation to Climate Change in Fishing Villages
        and Forest Settlements in Northwest Russia   227
        E. Carina H. Keskitalo and Antonina A. Kulyasova

12   Regional Governance, Path Dependency and Capacity-Building
        in International Environmental Cooperation   245
        Monica Tennberg

13   Summary   259
        Soili Nystén-Haarala

Index   *263*

List of Figures and Tables

Figures

I.I   Map of Northwest Russia   xvii

2.1   Factors Affecting the Forest Governance Regime in Russia   27

9.1   Relations Between Networks in a Market Campaign   173

9.2   Old Growth Forest Conflict Stages   181

Tables

7.1   Main Data Concerning the Localities Studied   135

7.2   Data on the Enterprises Studied   138

List of Contributors

Lyudmila Ivanova works as Senior Researcher and Head of the International Department of the Institute of Economics at the Kola Science Centre of the Russian Academy of Sciences in Apatity. She graduated from the Plekhanov Mining Institute in Leningrad in 1987 with a master's degree in the economics and organization of geological works. She has worked since that time as a researcher at the Kola Science Centre. Her PhD (Candidate of Sciences), completed in 2001, dealt with economics, planning and management of the economy. She has worked as a visiting researcher in Norway and Austria. Her main field of interest is forest management.

E. Carina H. Keskitalo works as Associate Professor of Political Sciences at the Department of Social and Economic Geography at the University of Umeå, Sweden. Ms Keskitalo has participated in several research projects at the Finnish and European levels, and is currently leading Swedish-funded projects on climate change adaptation in forest systems and on national adaptation to climate change. Her main areas of interest are international, northern and arctic cooperation and the impact of globalization and climate change on the regional level.

Jarmo Kortelainen works as Professor and Head of the Department of Geography at the University of Joensuu. He graduated in 1986 from the same department with a master's degree in human geography. He completed his licentiate degree in 1991 and doctorate in 1996. He has worked in a variety of research and teaching positions since 1985 and participated in a number of research projects. Mr Kortelainen's main research interests are mill communities and border studies.

Juha Kotilainen Dr. Juha Kotilainen is Senior Researcher at the Faculty of Social Sciences and Regional Studies at the University of Joensuu, Finland. With a background in human geography, his research interests include the analysis of the role of environmental issues in the transformation of the forest industrial sector, local developments in peripheral regions in Finland and Russia and the local perspective on cross-border interaction, one case study being the EU's external border between Finland and Russia. He has published in the journals *Environmental Politics*, *Eurasian Geography and Economics* and *Journal of Borderland Studies*, and recently edited the book (with Jarmo Kortelainen) *Contested Environments and Investments in Russian Woodland Communities*, published by Kikimora Publications. He is currently involved in the effort to merge the faculties of social sciences at the Universities of Joensuu and Kuopio (Finland), with a special focus on developing research on the topic of post-industrial challenges for resource-based communities.

Ivan P. Kulyasov is a researcher at the Centre for Independent Social Research in St. Petersburg. He graduated in 1991 from St. Petersburg State Medical Academy with a mater's degree in Medical Science. In 1997 he started to study ecological problems and social sciences and to work as a researcher at the Centre for Independent Social Research. Since 2002 he has participated in the postgraduate programme of the Department of Sociology at St. Petersburg State University. Mr Kulyasov has worked as a researcher in numerous international projects dealing with environmental issues and civil society. He has done a great deal of fieldwork in Russia and has thirty-eight publications.

Antonina A. Kulyasova is a researcher at the Centre for Independent Social Research in St. Petersburg. She graduated with a degree in Economics from St. Petersburg State University in 1991. In 2001 she received her PhD in Economics (Theoretical and Environmental Economics) from St. Petersburg State University. Since 1997, Ms Kulyasova has worked as a researcher and coordinator of the group for environmental research at the Centre for Independent Social Research. She has taken several international courses and worked in the United States and Finland on research grants. She has participated in numerous international research projects and has thirty-nine publications.

Anna-Maija Matilainen graduated in 2000 from the Faculty of Law of the University of Lapland, Finland, with an LLM degree. After graduation she worked for several years as a practicing lawyer and was trained on the bench in the District Court of Kuopio in 2001. In 2003, she started working as a researcher in a number of projects and studying for her doctoral degree at the University of Lapland. She worked in the project 'Governance of Renewable Natural Resources in Northwest Russia' in 2004 and continues to do research in the new project 'Trust in Finnish-Russian Forest Industry Business Relations'. Her article will form part of her PhD studies at the University of Lapland.

Soili Nystén-Haarala has been Professor of Civil Law in the Faculty of Law, Economics and Business Administration at the University of Joensuu since 2004 (tenured as of 2006). She completed her doctorate in the Faculty of Law of the University of Lapland in 1998. She has held a number of different positions at the University of Vaasa and University of Lapland. She has also worked as a visiting fellow in the Forestry Project at IIASA (International Institute of Applied Systems Analysis) and as a visiting professor at Umeå University. Ms Nystén-Haarala has led several research projects since 2003. Her main field of interest is contract law, international trade law and Russian law. In contract law she focuses on 'contracting', the use of contracts in business practice. She has publications on Russian law and transition as well as on contracting and proactive law. Among other projects, she has headed 'Governance of Renewable Natural Resources', in which all of the other authors have participated. Work in this field continues with the new project 'Trust in Finnish-Russian Forest Industry Business Relations', which is financed by the Academy of Finland.

Minna Pappila works as a researcher in the project 'Trust in Finnish-Russian Forest Industry Relations' at the University of Joensuu. She graduated in 1998 from the Faculty of Law of the University of Turku and received her licentiate degree in 2005. She has taken courses in environmental studies and participated in international study programmes. Her previous publications have focused on legislation, forestry and protection of biodiversity in Finland and Russia. Her article in this volume will form part of her doctoral dissertation at the University of Turku.

Svetlana S. Pchelkina works as a researcher at the Centre for Independent Social Research in St. Petersburg. She graduated from the Department of Journalism at St. Petersburg State University with a master's degree in 1989. Her principal academic interests are the environmental movement in Russia, forest governance and non-state governance of natural resources. She has participated in a number of international research projects.

Larissa Riabova works as Head of the Department of Social Politics in the North of the Institute for Economic Studies of the Kola Science Centre of the Russian Academy of Sciences. She graduated from Kiev National Economy Institute with a master's degree in 1982 and received her PhD (Candidate of Sciences) at the Leningrad Institute of Economy and Finance in 1989. She has held a variety of research positions at the Kola Science Centre since 1982 and worked as visiting researcher at the University of Lapland, Finland, University of Umeå, Sweden, and University of Tromso, Norway. She has participated in several international research projects and has an extensive list of publications. Her main field of interest is the economic and social impact of natural resource governance, in particular the governance of fisheries.

Monica Tennberg is Research Professor at the Arctic Centre of the University of Lapland. She graduated from the University of Helsinki in 1990 with a master's degree in Political Science, and received her licentiate degree (1994) and doctoral degree (1998) in Social Sciences (International Relations) from the University of Lapland. From 1992 to 2004 she worked as a researcher or as an assistant in the Department of Social Studies. Since 2004 she has been Research Professor in Sustainable Development at the Arctic Centre. Her main field of study is international environmental politics in the Arctic. She heads several research projects and supervises PhD students. Her most recent projects deal with indigenous identity politics, adaptation to impacts of climate change and Russian environmental politics.

Maria Tysiachniouk works as a researcher and the chair of the environmental sociology group at the Centre for Independent Social Research in St. Petersburg. She has a degree in Biology from St. Petersburg State University (1978), a master's degree in Environmental Studies from Bard College, New York (1998), a PhD

(Candidate of Sciences) in Biology from the Russian Academy of Sciences (1984) and is a PhD candidate in Sociology at Wageningen University, the Netherlands. Ms Tysiachniouk has taught courses in the United States at the Institute of Policy Studies at Johns Hopkins University and at Dickinson College, Townson University and Ramapo College. In Europe, she has taught at the Freiburg School of Forestry, Aleksanteri Institute and the University of Joensuu. She has done research in the United States on many fellowships, for example awards from the Fulbright Program, IREX and the Kennan Institute of Advanced Russian Studies. She has participated in numerous international research projects and has an extensive list of publications. Forest certification is her most recent research interest. Based on her research findings she also provides consulting services to logging companies and advises them on the social aspects of certification. She is currently a member of the Board of Directors of the International Sociological Association and a member of many other professional societies.

Stefan Walter was born in Germany, where he apprenticed in logistics and business management. He earned a Bachelor of Science degree in Geography and Environmental Management at Middlesex University, UK, in 2002 and a master's degree in Social Sciences (Sociology) in 2004 at the University of Lapland in Finland. Since 2004 he has been working towards his doctoral degree in the Department of Social Sciences, University of Lapland, and since 2005 has been a member of the Arctic Graduate School ARKTIS. His thesis will deal with societal capacity for sustainable management of renewable natural resources (forestry) with a focus on economic aspects. He participated in the project 'Governance of Renewable Natural Resources in Northwest Russia' as a researcher (7/2005 – 6/2006) and his article will be part of his doctoral thesis. He is currently a researcher in the Sustainable Development research group at the University of Lapland's Arctic Centre.

Preface

This book is based on the work of an international, multidisciplinary research group of scholars in the fields of human geography, environmental sociology, law, economics and international policy. Our research was made possible by the 'Russia in Flux' programme, which ran from 2004 to 2007 and was funded by the Academy of Finland (grant number 203964). The Academy also provided international exchange scholarships for members of the group, enabling them to conduct interviews and to create and maintain contacts with Russian universities, research units and cooperation partners.

The contributions in this volume began as cooperation between researchers from the Universities of Lapland and Joensuu and their international partners. The Russian researchers in the project come from the Centre for Independent Social Research in St. Petersburg and the Kola Science Centre of the Russian Academy of Sciences in Apatity.

Numerous people have contributed to the writing process of the book; acknowledging all of them here by name would produce too lengthy a list. Several Russian partners helped in arranging interviews and sometimes even participated in interviewing. Many scholars participated in the project seminars and provided fruitful comments on our ideas or read and commented on drafts of the articles. Our sincerest thanks go to all those who have played a part in creating this volume. Any errors, of course, remain the authors' responsibility.

Joensuu, 17 December, 2008

List of Abbreviations

AG	Aktiengesellchaft (German joint-stock company)
BEAC	Barents Euro-Arctic Council
CBC	Centre for Biodiversity Conservation
CBD	Convention on Biological Diversity
CBSS	Council of the Baltic Sea States
EEZ	Exclusive Economic Zone
ENA FLEG	Europe and North Asia Forest Law Enforcement and Governance
ENGO	Environmental non-governmental organization
EU	European Union
FFA	Federal Fisheries Agency
FSC	Forest Stewardship Council (certificate system)
FZ	Federal'nyi zakon (Russian Federal Law, act of parliament)
GmbH	Gesellschaft mit beschränkter Haftung (German limited liability company)
IES	Institute for Environment and Sustainability
IFF	Forum on Forests (now UNFF)
IMF	International Monetary Fund
IPF	Intergovernmental Panel on Forests (now UNFF)
KSC	Kola Science Centre
MMBI	Murmansk Marine Biological Institute
MNR	Ministry of Natural Resources (of the Russian Federation)
NDEP	Northern Dimension Environmental Partnership
NEFCO	Nordic Environmental Finance Corporation
NGO	Non-governmental organization
NIBR	Norwegian Institute for Urban and Regional Research
NPM	New Public Management
OAO	Otkrytoe aktsionernoe obshchestvo (Russian open joint-stock company)
OECD	Organization for Economic Cooperation and Development

PINRO   Polar Research Institute of Marine Fisheries and Oceanography

RAS   Russian Academy of Sciences
RF   Russian Federation
RNR   Renewable natural resources

SEU   Socio-Ecological Union
SFM   Sustainable forest management

TAC   Total available catches
TACIS   Technical Assistance to the Commonwealth of Independent States
           and Mongolia (EU Programme)
TC   Transnational corporation

UN   United Nations
UNESCO   United Nations Educational, Scientific and Cultural Organization
UNFF   United Nations Forum on Forests
US   United States (of America)
USSR   Union of Soviet Socialist Republics

WTO   World Trade Organization
WWF   World Wide Fund for Nature (formerly World Wildlife Fund)

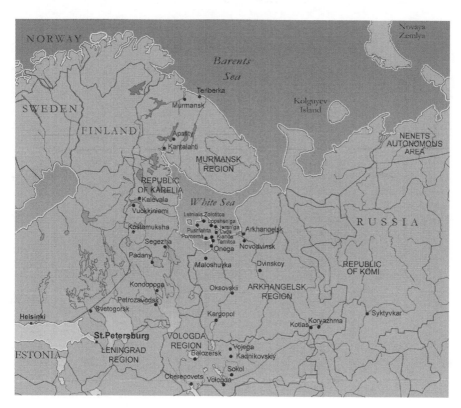

Figure I.I   Map of Northwest Russia

Chapter 1
Introduction

Soili Nystén-Haarala

Research focus

This book is based on the work of an international, multidisciplinary research group studying transition in Russia. The focus is on the adjustment of local communities and enterprises to institutional changes, as well as their attempts to govern that development. Of particular interest is the governance of forest and fishery resources in Northwest Russia. The local view is approached empirically, with data gathered through interviews of local authorities, managers and people involved in the governance and use of natural resources. These 'empirical local views' are examined against the changes in official institutions on the national level and in the global arena in order to elucidate the interplay of official and unofficial institutions.

Northwest Russia

Northwest Russia can be understood geographically, administratively, politically or economically. Geographically the European part of Russia bordering Fennoscandia and the Baltic Sea has been called Northwest Russia. In Russia itself, the term 'Northwest Russia' was seldom used and did not mean anything in particular before President Putin's decree in 2000 creating seven new administrative areas (*okrug*) under presidential administration.[1] One of these areas was the part of Northwest Russian covering the Karelian Republic, the Murmansk, Arkhangel'sk, Komi, Leningrad, Novgorod, Vologda, Pskov and Kalingrad Regions and the city of St. Petersburg (see the map at the beginning of the book).

The Russian Constitution divides the Federation into 89 'subjects' (*sub"ekt federatsii*), which we refer to as 'regions' in this volume. According to the Constitution, the regions are equal. However, they comprise different subdivisions: republics (*respublika*), regions (*oblast'*), territories (*kray*), cities of federal importance (Moscow and St. Petersburg), (the Jewish) autonomous region

---

1 These areas were placed under administration by the prime minister when the new president, D. Medvedev, appointed Putin as his prime minister.

(*avtonomnaya oblast'*) and autonomous areas (*avtonomnyi okrug*). Autonomous areas are often situated within the borders of another region.[2]

In this book we use the term 'Northwest Russia' in its geographical meaning, focusing mostly on the areas bordering Fennoscandia and the Baltic Sea. Economically this area is quite heterogeneous. The forest economy is important in most parts of the region. Russia has 809 million hectares of forests – more than one-fifth of the forested area in the world. However, in a country with huge resources of oil, natural gas and different minerals, including diamonds, the forests have not been the natural resources of greatest interest to the federal government. Murmansk is less dependent on forestry and more dependent on heavy industry (the mining, metal and machinery industries), sea transportation and large-scale fishery. The Karelian Republic and the Arkhangel'sk Region are both covered by forests and dependent on them. Arkhangel'sk, however, also has a great deal of machinery and other heavy industry, as well as transportation and sea fishery, whereas Karelia, while it also has mining, metal and machinery industry, is mostly dependent on its forest resources and forest industry. The Stockman oil field in the Barents Sea and the future pipeline to Murmansk harbour will also bring the oil industry into the area. At present, industry in Arkhangel'sk, Karelia and Murmansk is quite dependent on coal in this oil and natural gas rich country.

The interdependence of different parts of Northwest Russia is quite weak compared to each region's political and economic dependence on the federal centre (Moscow). The economic influence of St. Petersburg is important, especially in the surrounding Leningrad region, which has attracted a huge amount of international investment, for example, the Svetogorsk pulp and paper mill (International Paper), a Ford Motors car factory and a Nokia Tyres plant. The reasons for the amount of international investment can be found both in the region's location near St. Petersburg and in its favourable investment policy. The economic importance of St. Petersburg is growing also further away in the Karelian Republic and Arkhangel'sk Region. The economic growth of the city of St. Petersburg is certainly unique in Northwest Russia, but it falls outside the scope of the book.

The Research Process

The authors of the book comprise an international multidisciplinary research whose work was financed by the Academy of Finland's programme 'Russia in Flux' (2004–2007; number 203964). The common theoretical framework of the

---

2   President Putin started a process to reduce the number of regions (subjects). This currently affects several autonomous areas, which are to be made part of the regions (oblast or republic) within which they are located. Since all of the regions are enumerated in the Constitution, such changes require referendums in the regions concerned as well a law which has to be passed by a two-thirds majority in both chambers of the Federal Assembly. Two such laws have already been passed.

group, which consists of scholars in the fields of human geography, environmental sociology, law, economics and international policy from Finland, Russia, Sweden and Germany, is based in Douglass C. North's concept of path dependency. The group emphasizes the path-shaping elements associated with path dependency rather than its more deterministic elements in order to stress the active influence of and interplay among different interest groups. The project was directed by Soili Nystén-Haarala, editor of the present volume.

The empirical data for the research, mainly interviews, were collected in the Karelian Republic, the Arkhangel'sk, Murmansk, Vologda and Leningrad Regions and in the cities of Moscow and St. Petersburg between the years 2004 and 2007. Maria Tysiachniouk, Antonina Kulyasova and Ivan Kulyasov collected most of the empirical data. Some interview trips into remote fishing villages of the Arkhangel'sk Region and forestry locations in the Vologda Region took several weeks or even months. Soili Nystén-Haarala and Anna-Maija Matilainen conducted interviews in Arkhangel'sk, Murmansk, Petrozavodsk, Kondopoga and Segezha. Larissa Riabova and Lyudmila Ivanova collected data in Murmansk and municipalities in the Murmansk region. The interviewees comprised federal, regional or local authorities, company managers, workers, local inhabitants and NGO members, all of whom were involved in the governance and use of natural resources. Monica Tennberg's interviews, conducted with the assistance of Tamara Semenova from the Russian Institute of Cultural and Natural Heritage of the Russian Academy of Sciences, Svetlana Agarkova from Petrozavodsk State University, Nadezhda Kharlampieva from St. Petersburg State Univertsity, Larissa Riabova and Lyudmila Ivanova from Kola Science Center of the Russian Academy of Sciences and Antonina Kulyasova from the Center for Independent Social Research, focused on people who had participated in international environmental projects in Northwest Russia. The interview locations are shown on the map at the front of the book.

The research group have written the contributions in close cooperation. The first planning meeting was held in Rovaniemi, Finland, in spring 2004. The drafts were commented on in seminars held in September 2005 in St. Petersburg, in spring 2006 at the Pyhätunturi resort in northern Finland and in September 2006 in St. Petersburg.

Structure of the Book

National Policies and the Transition to a Market Economy

The book consists of three different parts. It starts with an analysis of national policies and their impact on Russia's transition to a market economy, with a particular focus on how the institutional framework has developed. Chapter 2 is an introduction to the theme of the book and analyzes general tendencies in the transformation of Russian institutions and the interest groups that have endeavoured to influence the development of those institutions. The contribution also introduces the concepts of governance and path dependency. It analyzes and questions the path dependency of the developments studied and briefly evaluates the theoretical approaches of institutional economics.

In Chapter 3, researcher Stefan Walter analyzes fiscal and monetary policies and their effect on economic development in Russia. His focus on the trust-maintaining function of macroeconomic policy helps illuminate the mutual dependency of good economic policy and forest resource governance. Combining economics and sociology, Mr Walter applies Niklas Luhmann's theory of social systems. The chapter is a part of Mr Walter's doctoral studies in sociology at the University of Lapland.

In Chapter 4, researcher Minna Pappila analyzes the drafting process of the Russian Federal Forest Code which came into force at the beginning of 2007. She focuses on sustainable development and the difficulty of its integration into the new code. In spite of severe criticism by environmental NGOs of the drafting process and the content of the Forest Code, the legislation can be regarded as reflecting the development of democratic discussion and the participation of civil society in law drafting. Ms Pappila's contribution is a part of her doctoral thesis in the Faculty of Law at the University of Turku.

Dr Larissa Riabova and Dr Lyudmila Ivanova write about the painful changes in fishery management in Northwest Russia in Chapter 5. The focus is on the development of the governance framework, which in Russia, too, is proceeding from government to governance. The authors claim that the power struggle between the federal and regional levels, as well as the emergence of municipal self-government, is the starting point for decentralization and the first sign of a future development towards multilevel governance in Russia, where the dominant role of the state has deep historical roots.

In Chapter 6, researcher Anna-Maija Matilainen describes the takeover struggles for companies within the forest industry. She analyzes how company law in the Western market economy has been used – or, rather, misused – in a creative way in the struggle to gain ownership of companies. The legal analysis of hostile takeovers reveals an economic and business environment that is peculiar to Russia; Ms Matilainen examines the terms on which this environment operates with reference to both her own interviews and articles in local and national newspapers reporting takeovers. Ms Matilainen's contribution is part of her doctoral studies in the Faculty of Law at the University of Lapland.

Case Studies on Different Aspects of Governance

The second part of the book consists of case studies illustrating different aspects of governance. In Chapter 7, Dr Juha Kotilainen, Dr Antonina Kulyasova and researchers Ivan Kulyasov and Svetlana Pchelkina focus on the interaction of different interest groups in forest governance, which officially is strictly governed by the state. The empirical data are largely based on numerous interviews of local managers of logging companies, directors of municipalities and workers and residents of logging settlements in the Arkhangel'sk and Vologda Regions and the Karelian Republic. The chapter illustrates an interesting process of hybrid governance in the form of interaction between official and unofficial institutions.

Professors Soili Nystén-Haarala and Jarmo Kortelainen of the University of Joensuu analyze in Chapter 8 how trust is constructed and maintained in Russian mill towns. The chapter compares different forms of trust building and governing relations between pulp and paper mills and their surrounding communities. Community relations are important in the social and economic circumstances of Russian mill towns, where the mill management faces challenges in balancing between economic and social requirements.

In Chapter 9, Dr Maria Tysiachniouk analyzes how Greenpeace caused and used conflict in the Karelian Republic by informing Western European consumers of the logging of old-growth forests and lobbying for a project that sought to establish a national park in the republic. The conflict and its results are studied based on interviews of local residents who make their living from forest resources in one way or another and representatives of the environmental movement. The contribution analyzes how civil society may break into the field of traditional state governance and uses international consumer boycotts as weapons to achieve its own goals in nature protection.

International and Global Impacts on National Environmental Policy and Local Forestry and Fishery

The third part of the book raises the discussion from empirical case studies to the international level, focusing on international and global impacts on Russian environmental policy. Both contributions are based on analyses of interviews.

Chapter 10 is a collaborative contribution written by Associate Professor Carina Keskitalo and researcher Antonina Kulyasova. Dr Keskitalo has previously done research on the vulnerability of forest resources, fisheries and reindeer herding to processes of globalization and climate change in the Scandinavian north. Ms Kulyasova conducted interviews in fishing villages and forest localities of Northwest Russia. Their research shows the similarity of problems in northern areas that are dependent on renewable natural resources as well as a number of differences, caused by the earlier socialist economy in Russia and the transition to a market economy.

In Chapter 11, Professor Monica Tennberg focuses on international environmental co-operation in Northwest Russia. She interviewed both Russian and Finnish participants in environmental projects financed by the European Union and analyzes whether and how they have managed to build local capacity. The chapter describes how participants from different sides of the border experienced cooperation and how they viewed the results of the projects not only as governance but also as capacity building for the project participants.

Chapters 3 and 8 were already written in 2006. Most of the authors have worked on their contributions and updated them to May 2008. In some cases, developments in the forestry and fishery sectors of Northwest Russia in June 2008 have been included as well. Political developments following the parliamentary (Duma) elections of 2007 and the presidential elections of March 2008 have not been analyzed in this volume.

PART I
National Policies and a Transition to a Market Economy

Chapter 2

Institutions, Interest Groups and Governance of Natural Resources in Russia

Soili Nystén-Haarala and Juha Kotilainen

Introduction

The collapse of the Soviet Union was a profound change – one that reshaped the world both politically and economically in the last decades of the twentieth century. Mikhail Gorbachev's *perestroika* and *glasnost'* aimed at changing the planned economy gradually under the rule of the Communist Party, while the reformers – with Boris Yeltsin as their figurehead – wanted to introduce a market economy, democracy and the rule of law immediately through sweeping changes. The demise of the Soviet Union opened up a new path for the development of Russia and started a transition process that has seen an interesting interplay of global and local impacts on Russian national, regional and local institutions.

In the first shock of rapid privatization, marked by the disappearance of state property and the shaking of the legal order, the state lost its traditionally overwhelming power, a new business elite emerged and foreign governments and international organizations started to aid and advise the Russian government. The economic and political ideals that dominated at the time affected the Russian transition, but they also encountered Russian institutions imbued with totally different ideals, working methods and mindsets than their Western counterparts. Old and new mixed in a unique way. After about twenty years of transition, many scholars call the outcome of this transformation process 'Russian democracy', 'the Russian market economy' and 'the Russian rule of law' (see e.g. Oleinik 2001; Sutela 2003). Russia may not even be heading towards an ideal Western model, but producing its own variant with a strong Russian flavour.

The first Russian reformers and their Western advisers paid no attention at all to institutions (see e.g. Åslund 1995 and Sachs 1993), because the then dominant neo-classical economics dealt only with stable systems without analysing change (North 1990). According to North,

> [i]nstitutions are the rules of the game in a society, or more formally, are the humanly devised constraints that shape human interaction. In consequence they structure incentives in human exchange, whether political, social, or economic. Institutional change shapes the way societies evolve through time and hence is the key to understanding historical change (p. 3).

Russia and other transition economies have been an interesting laboratory for understanding how significant a role institutions play in change.

Legal studies, which contribute to the change of official institutions, are also based on a static approach to the existing legal system and see change in technical terms, focusing only on the legal framework and the legal system. However, the new institutionalism asserts that official institutions (formal constraints), such as legislation, economic rules or formal contracts, can technically be changed quite quickly, but unofficial institutions (informal constraints), such as attitudes, working habits and behavioural rules, change much more slowly and prevent official institutions from changing in the planned, 'ideal' direction. One constant challenge in the Russian transition has been that unofficial, informal institutions do not support the official ones. According to institutional economics the interplay of formal and informal rules is the key to understanding economic development. (North 1990)

In this chapter we present a framework of institutions and interest groups and discuss their interplay, which illuminates the decisions and solutions made at the local level in Northwest Russia. We first present governance and path dependency, the main concepts applied by the authors of this volume in their empirical studies of transformation at the local level. Secondly, we present the main actors and interest groups: the state, municipalities, private companies and NGOs with the framework of official institutions. Thirdly, we show how the interplay of official and unofficial institutions is developing and what role different interest groups play in this development.

The Main Concepts

Governance

Governance, although derived from the word 'govern', is a broad and multidisciplinary concept, implying something beyond government administration. Kersbergen and van Waarden (2004) have classified nine different approaches to and meanings of governance, giving a complex view of the multidisciplinary studies of governance. We present them all here briefly, because the authors of this volume, who represent a variety of disciplines, use the concept in several of these meanings.

Economic governance is usually connected with private governance and often seen as a mechanism that firms choose in order to save transaction costs and survive on the market (Williamson 1985). As a concept connected with institutional economics, economic governance has been developed within different disciplines, including economic history, economic sociology and political economy. Approaches based on economic governance view markets as being created and maintained by institutions. Governments are regarded as only one source of institutions, and private governance is seen as a more effective way to solve problems than state governance. This way of thinking can be traced back to the US understanding, inherited from colonial times, that the role of the state should be minimal and that the state is not to be trusted (see e.g. Nozick 1974).

In Russia private governance could be a path-breaking way out of the overwhelming but ineffective state governance that has a long and strong tradition in the country. However, the attitude towards the role of the state is very far from that of Anglo-American political libertarianism. The reformers of the beginning of the 1990s were convinced of the need to diminish the role of the state, but since the beginning of Vladimir Putin's presidency the Russian political elite has become more interested in restoring a stronger role for state governance and convinced of the need for a strong state, which for them represents order.

In the international arena, organizations such as the World Bank or the IMF have started to use the concept *good governance* to refer to both government and non-government economic policy in their reports. Good governance is promoted by these organizations and this often means government implementing neo-liberal policy by reducing public spending, privatizing state enterprises and the banking sector (*the Washington consensus*)[1] or, nowadays, also by promoting greater transparency and accountability in public and corporate affairs. After the scandals with misuse of financial aid in Russia,[2] these organizations have given combating corruption a very prominent place on their agendas. The OECD has propagated good governance by comparing best practices in key areas such as public management, business–government relations and social policy. In the beginning of the Russian transition, the World Bank and the IMF were quite influential and could direct Russian economic policy through the conditions for international economic aid; nowadays, however, Russia's official policy is to restore the national pride of the collapsed superpower, distance the country from international organizations and foreign governments and make independent decisions.

The term 'good governance' is also used in the private sector in the meaning of *corporate governance*, emphasizing accountability and the transparency of the actions of management towards shareholders or even a broader circle of stakeholders. The OECD has established a set of non-binding principles of corporate governance that represents a common basis that OECD member countries consider essential for the development of good governance practice (www.oecd.org/daf/governance/ principles.htm). Good corporate governance is assumed to benefit government in the form of increasing international investments. In Russia, corporate governance is seen from another perspective. Potential conflicts of interest between shareholders and management are not the most important issue for Russian companies; what is crucial are social relations with the state, the local community and their workers. Transparency is not one of the main virtues of Russian markets, where harsh means to gain profits and to beat competitors are rather common. According to

---

1   For criticism of the Washington consensus, see Stiglitz 2002.

2   Stiglitz 2002, 133–165. Before the aid programmes for post-socialist countries, misuse of financial aid was ignored or at least not discussed openly in international organizations.

some Russian authors, the global economy constrains the Russian economy and will inevitably lead to the need to follow good governance.[3]

The *New Public Management* (NPM) literature has also introduced good governance in public organizations. For the most part, NPM focuses on the similarities between public sector reforms in countries that are politically and economically very different (Kersberger – van Waarden 2004). In all these reforms, the market is a model for public policy implementation. NPM is inspired by public choice, principal-agent theory and transaction cost economics (Kaabolian 1998).

In international relations, governance is often used to describe *governance without the government* (e.g. Rosenaur and Chempiel 1993). International politics and international law are traditionally arenas dominated by governments without any organization above them, since the powers of international organizations derive from the states that comprise them. The sovereignty of the state was strongly emphasized in the Soviet Union, which did not accept any organization or principle above the state (see e.g. Långström 2003). Nowadays international law and politics produce an increasing number of treaties, commitments and organizations to guide governments. For example, international treaties such as the Kyoto Protocol have created control mechanisms that are above governments.

The development from government to governance is seen as a new phase of modernity and is connected with the diminishing role of nation-states, international networking of private actors and networking without the government playing a leading role(e.g. Magone 2007; Hoogle and Marks 2003) R.A.W. Rhodes defines governance as self-organizing interorganizational networks (Rhodes 1997). The EU is nowadays often characterized as a *multilevel governance system*, in which the union (both political and legal), government and private levels interact. According to Hooghe and Marks, structural policies of the 1990s led to self-organizing networks among different levels of government. The diversity of the territorial political structure of the EU countries created a heterogeneous system of self-organizing international networks which constitute EU multilevel governance (Hooghe and Marks 2003).

In Russia, with its ideal of strong centralized state power inherited from Soviet and even earlier times, transition started a struggle to implement federalism and introduce local self-governance. As several authors in this volume suggest (in Chapters 5, 7, 9 and 12), governance is also taking on new forms in Russia and extends from the traditional state governance (government) to cooperation between business and NGOs without the strong guidance of state power.

The concept of governance is also used to describe *self-governance*. An example can be found in Elinor Ostrom's research (1990) on governance of the commons for common local needs and governance that is independent of a government. Similar forms of self-governance can be found in the Russian countryside, where they have evolved because the central power – which in the Russian conception

---

3   For example, Radaev's studies on the gradual legalization of Russian business suggest that such a change is on the way, although there are obstacles which should not be ignored or underestimated (Radaev 2002).

should be responsible for everything in the country and which during Soviet times tried to crush all self-governance – has actually rejected or forgotten remote areas, leaving them to survive on their own. Antonina Kulyasova and Ivan Kulyasov's case studies on remote fishing villages inhabited by Pomors in the Arkhangelsk Region focus on this issue (see Chapter 10).

Many researchers have pointed out the difficulties of dealing with the new concept of governance (e.g. Rhodes 1997). Shifts in governance reflect changes in the development of societies. Our Western political system is based ideologically on democracy, which is built on the institutions of nation-states. Development of modernization or post-modernization has weakened the nation-state. Supranational power and especially the power of the global market forces have constrained and diminished the traditional democratic decision-making power of nation-states. The old checks and balances do not work efficiently anymore. A shift of power to specialized agencies, international organizations or private companies diminishes the accountability of political power and poses the threat of a legitimacy crisis.

Attempts have been made to overcome the deficit in democracy by handing decision-making power to judicial organs, which often reluctantly have to make decisions where politicians are incapable of doing so. The EU is a good example of the power of judicial organs in developing and shaping integration. The enlargement of markets has forced companies to network and integrate with each other and with public power. Governability might increase in this way, but accountability becomes more problematic. Markets govern private companies, but the legitimacy of the public sector cannot be governed simply by markets. Democratic accountability and legitimacy are in crisis and new solutions are being sought in the Western world (Kersberger – van Warden 2004).

It is evident that the democracy and market economy ideology that were introduced in the post-socialist countries was itself in crisis and in need of new, more creative paths and solutions. An idealistic 'good governance' model does not fail only because of different economic and political circumstances but also because it is an ideal that no longer works properly in old, advanced democracies either. The authors of this volume analyze existing forms of governance, which often are attempts to adjust to new circumstances and which mix old informal institutions and new formal institutions of Western models. The change is in any case institutional and proceeds through institutions. As North puts it:

> We are still in the beginning to understand change of institutions. Analysis of the interplay between formal and informal institutions in transforming societies can help us to find new answers to how economy and society change (North 2005, p. 64).

Path Dependency

When old and new mix, the concept of path dependency takes on importance. Path dependency can be understood in different ways. Along with helping us to understand that 'a specific developmental path tends to be re-established once it has been instituted even if it no longer is optimal' (North 1990), it can also strengthen a determinist attitude that hinders path breaking and path shaping. Path breaking may seem to appear impossible because of strong embedded forces that drive path dependency. However, according to North, path dependency only helps us to understand that history matters (North 1990).[4] Yet the strength of the past should not be overestimated with this concept. On the other hand, a path-dependency approach implies that social forces can intervene, correct and actively shape developmental paths so that new orientations become possible (Nielsen-Jessop-Hausner 1995). Transition is a process that aims at path breaking and path shaping. There are also many social forces pushing for path breaking. Conscious path shaping requires an analysis of the interplay between the official and unofficial rules of the game.

Institutions matter on different levels of society. Oliver E. Williamson's presentation of four different levels of institutions illustrates the complicated interplay between the levels (Williamson 2000). First, there is the embedded level, which consists of cultural differences and ways of thinking (see Granovetter 1985). These can most often be traced back to religion and culture. This level sustains path dependency unconsciously and therefore most effectively. Legislation and other official rules are the second level of social institutions. Lawyers and social scientists, who work on this second level, are interested in putting the framework right.

Williamson distinguishes a third level of institutions for the behaviour of companies in particular. Firms have to choose the best possible mechanisms of governance, as these save transaction costs and function in the most profitable way. According to Williamson, firms choose between markets and hierarchies (organizations) and hybrids between them. The governance of firms is a process of adjustment to framework conditions. As an interest group, firm managers definitely also try to contribute to creating better framework conditions for their firms, which makes them lobby for changing the rules of level two.

The fourth level comprises the day-to-day decision making of firms and other actors. Everyday routines are also institutions and evolve and change in the framework that the institutions on the three other levels offer them. The authors of this volume focus on the interplay of institutions at different levels. For instance, legislation is studied as the interplay of different interests during the drafting process (Chapter 4) and as a tool in corporate takeovers (Chapter 6). Our empirical studies focus mostly on the governance methods of firms and other actors at the local level. Their solutions can be path dependent, or quite often also path breaking – especially when actors representing different interest groups consciously try to find new forms of cooperation applying both official and unofficial institutions.

---

4   North discovered the idea of path dependency from the work of researchers studying technological change. He mentions David and Arthur. According to North, Arthur studied decision making in firms, which North then elaborated into an institutional model (North 2005).

The strength of path dependency is illustrated in many studies of Russian transition. The importance of informal institutions is often regarded as especially strong in Russia. Richard Rose explains this phenomenon as a sign of an *antimodern* society where official institutions have failed, leaving the citizens no other alternative than unofficial institutions such as personal networks, influencing the sympathies of civil servants and the last resort, corruption (Rose 2000). The official institutions of a modern (not post-modern) society are based on (Weberian) formal rationality, which seeks to guarantee the same result in similar circumstances by strict application of formal rules. While Western countries have struggled with post-modern governance since the 1970s (Magone 2006), Russia has never really gone through this basic phase of modern society.[5] On the contrary, informal institutions – especially networks between people – helped Russian people through the hard times after the collapse of the Soviet Union and the economic reform (see e.g. Kosonen 2002, Lonkila 1999).

An antimodern society has all the official institutions of a modern society, but these institutions do not function properly. Russian historian and political scientist Dmitriy Furman explains this phenomenon forming the gap between democratic legislation and real practice of politicians as an *imitation of Western institutions*. Official institutions which have been brought in from the outside have not been internalized. Furman claims that Western democratic institutions such as free elections, a multiparty system and the separation of powers technically exist in Russia, but in practice they are more of a façade and do not represent these institutions in the proper sense of the word (Furman 2007). The Russian political elite actually reinforces this dissemblance with an embellished concept of *managed democracy*, which suggests that the Russian people are not ready for democracy. In reality, the concept of managed democracy covers up the Russian political elite's fear of losing its power.

Looked at from the local level, the situation does not appear to be as grim as might be suggested by the approaches emphasizing the strength of informal institutions and harmful path dependency as well as the difficulty of introducing post-modern institutions in Russia. The interplay of official and unofficial institutions (be it antimodern or not) has resulted in a new and unique functioning of institutions. Organizations will evolve to take advantage of opportunities which the framework set can offer. There will be coordination effects via contracts with other organizations. Even more importantly, the formal rules will result in the creation of a range of informal rules that modify the formal ones and extend them to a variety of specific applications. The independent web of an institutional matrix produces massive increasing returns (North 2005).

---

5   Formal rationalism is often seen as a starting level for developing the rule of law. The idea can be traced back to Max Weber's presentation of formal rationality (see Weber 1972, *Wirtschaft und Gesellschaft* (Tübingen: J.C.B. Mohr)). Weber saw excessively developed formal rationality as a threat to freedom in society.

With increasing returns, institutions can shape the long-term path of economies as long as the markets are competitive and development is effective. But if the markets are incomplete, information feedback fragmentary and transaction costs significant, the subjective models of actors will shape the path; not only can poor performance then prevail, but historically derived perceptions of the actors may shape the choices that they make. A complex environment constrains the actors and prompts them to find help in available mental constructs – ideas, ideologies and theories (North 1990, 95–96).

New solutions that combine old and new approaches at the local level can prove to be ineffective and path dependent, but they may also break paths and open up new opportunities, serving as good examples for Western models struggling with their own accountability v. governability problems. The problem of imperfect markets should not, however, be forgotten, and it reminds us of the great importance of the policies of governments as well as international cooperation in providing actors at the grassroots level with effective means for solving their own problems.

Actors and Interest Groups in Natural Resource Governance

The Structure of the Federation and the Role of the State

Historically, the model for Russian rulers has been a strong state. The difference between Western and Eastern development started already during the Middle Ages, when the Catholic Church in the West created the embryo of a modern state based on law and exercised religious power while the princes took care of secular matters. The Orthodox Church, however, carried on the Byzantine concept of power, in which the tsar was the head of both the Church and the state (see e.g. Berman 1983). The Soviet regime modernized (industrialized) Russia with harsh methods, but continued to promote a uniform concept of power.

Even though the reformers of the 1990s admired libertarian economics, which is based on the American ideal of a minimal state, the great majority of Russians still believed in the virtue of a strong state, as their history books had explained. President Putin, whose background – like that of most of the political elite – is in secret service agencies (see Krystanovskaya 2004), systematically started to strengthen the state, 'taking back' the power which had fled into new business circles and the regions.

In the Soviet Union, the state had taken care of everything. The planned economy was an attempt to abolish markets. The idea of a strong state also included a belief in the virtue of centralization and a clear chain of command. The Soviet Union was a federal state, but only in name. The legislation of each republic was identical word for word. In practice, however, the centre could not control everything. The more remote the republic, the more the Communist Party leader of the republic was able to rule as a 'vassal' – as long as he or she reported that the state plans had been fulfilled (Hosking 1985).

The reformers, who came into power with President Yeltsin, believed in decentralization, counting on the fact that when the regions could start competing like the states of the US or the EU, the economy would develop in a more effective direction. The regions regarded the ownership and control of natural resources as being the core issue of decentralization and development of federalism in Russia. To a certain extent, decentralization was also a means to buy the support of the regions for Yeltsin in the struggle for power between the president and the parliament in the federal centre. When the president won the struggle, he started to centralize power again. Decentralization had not led to effective competition between the regions, but created centrifugal tendencies which threatened to tear the whole Federation apart (Nystén-Haarala 2001).

After becoming president, Putin continued centralization successfully, calling it a strengthening of the 'vertical power'. Again property rights and the income from natural resources were the core issue. Yeltsin did not succeed in making the appointment of governors of the regions one of the powers of the federal president; they started to be elected in direct elections in the regions. After that, he appointed his own representative for each region to see to it that federal legislation was implemented there. The regions, however, refused to obey. Many regions had passed their own legislation on natural resources, declaring them to under the ownership and control of the region. Putin managed to remedy Yeltsin's failure with a decree[6] dividing the 89 regions (subjects of the federation) into seven districts (*okrug*) under presidential administration and appointing people with military or secret service backgrounds to head these districts. One of these districts is Northwest Russia (see Chapter 1 of this volume). In this way, without any change in the Constitution, that is, by presidential decree only, Putin managed in practice to centralize the government of the regions under his own administration.[7] Later he also ousted governors from the Federal Council, the other chamber of the Federal Assembly (the parliament), where they represented the executive power of their regions.[8] Now the federal president appoints the heads of the executive branch, who in turn appoint the representatives of the executive branch to the Federal Council. The representatives of the legislative branch are elected by regional parliaments. The present federal legislation makes Russia appear more like a unified state than a federation.

---

6   Dated 13 May 2000, No 849.

7   The presidential administration is powerful in Russia and it has been compared with the earlier Communist Party administration. None of the other state organs has an administration of comparable size. Putin seems to want to take a part of this administration including the control of the seven districts (*okrug*) with him in his new position as prime minister. In this way executive power will again be divided in a new way without changing the Constitution.

8   According to article 95.2 of the Federal Constitution, each region has two representatives in the Federal Council, one representing the legislative organ and the other the executive organ of the region. The Federal Council can block any law with a majority vote, which the Duma can override with a two-thirds majority (article 105). Some laws, such as the federal budget, have to be passed by the Federal Council (article 106).

The division of powers between the Federation and the regions is regulated in articles 71–73 of the Federal Constitution. Article 71 lists the powers of the Federation, article 72 declares which powers fall under the joint jurisdiction of the Federation and the regions, and article 73 declares that all the other powers fall under the jurisdiction of the regions. All commercial and almost all civil legislation is completely federal, while regulation of natural resources, health, educational and social issues comes under joint jurisdiction. However, the Constitution does not regulate what joint jurisdiction actually means. The unclear division of powers between the Federation and the regions was one of the disputed issues in the drafting of the Constitution of 1993. Regional leaders were not satisfied with the ambiguity of the Constitution.[9]

During Putin's presidency, the provisions of the Constitution were made more precise through federal legislation that set out the principles of joint jurisdiction in detail that did not leave much for the regions. Establishing 'vertical power' became possible with the strong power of the federal president. Already Yeltsin's[10] constitution gave the president the power to initiate laws, sign and promulgate the laws accepted by the Federal Assembly and issue decrees. Yeltsin ruled with decrees, because the Duma could not agree on new federal laws or he himself blocked new legislation. Putin, however, managed to start a huge wave of new federal legislation, including long-disputed legislation on natural resources,[11] by unifying the Duma. This happened by establishing a party supporting the policies of the president and the government.

The map of Russian parties was rather mixed and parties were gathered around charismatic people. Putin changed the whole map by creating 'the power party', although during his presidency he was not even a member of it. The Unity Party

---

9   According to the presidential decree on referendum of October 1993, the Constitution had to be supported by more than half of the voters and more than half of those who were eligible to vote had to go to the polls. Many governors, however, encouraged the citizens of their regions to boycott the referendum. The Central Election Committee announced that 56.63 per cent of those eligible to vote participated in the elections and that 56.8 per cent of that number voted in favour of the Constitution. Later there were claims that millions of extra votes had been added to the results. The Central Election Committee 'destroyed the evidence' so that these claims could not be investigated. Wild rumours continue (see Nystén-Haarala 2001, 26-28).

10   The Constitution of 1993 was called 'Yeltsin's constitution', because he participated in drafting it personally, for example, by leading some meetings of the drafting committee and influencing the content, especially where the strong power of the president was concerned. He also put effort into pushing the Constitution through all the obstacles.

11   New legislation on natural resources started with passing of a new land code (dated 25.10.2001). This was followed in 2002 by amendments to the Law on Subsoil Resources of 1992 that gave the decision-making power on income from such resources to the federation. The new Forest Code of 2006 and the Water Code of 2006 are also outcomes of this productive wave of new legislation. Putin's presidency can be assessed as extremely efficient in shaping official institutions.

(later United Russia) became the means to acquire power and create a career for ambitious politicians much in the way the Communist Party earlier represented the path to the top of society. Deputies of other parties joined this new party, which immediately won the elections of 1999.[12] With both chambers of the Federal Assembly under his control, Putin could rule the country. Regional power was also diminished by abolishing regional parties and allowing only parties with federal importance.

During Putin's presidency, the constitutions of the regions and their legislation were put in line with the federal models. Differences between several regional constitutions, mostly regarding the ownership of natural resources, were taken before the Constitutional Court, which declared them to be in violation of the Constitution.[13] Treaties that President Yeltsin had concluded with heads of the regions giving regions more powers than the federal Constitution allowed were not continued. The asymmetry of the Federation, which many influential Moscow-based constitutional lawyers criticized, was turned into symmetry under the federal order. (See e.g. Shulzhenko 1995).

Article 10 of the Russian Federal Constitution sets out the doctrine of separation of powers. The Duma has the legislative power, the president and the government the executive power, and the courts the judicial power. The president also has considerable legislative powers – the power to issue decrees and to refuse to sign and promulgate the federal acts of the Duma. Parliamentarism, however, is quite weak even according to the Constitution, because when the Duma gives a vote of no confidence to the government, the president can dissolve the Duma (Article 117.3). The president can also dissolve the Duma (article 111) where it disagrees with the president's choice for prime minister.[14] The government is a tool by which the president can rule the country.

Judicial governance of the courts in the political system was introduced in Russia with the Constitutional Court. The Court has tried to interpret the Constitution in a formal way, emphasizing its wording yet realizing the political aspects of its decisions. In the centralization process, the Constitutional Court has supported the federal centre, interpreting the Constitution according to its wording. It has declared that regional constitutions violate the Federal Constitution (see

---

12   In 1999 the Unity Party became the second largest party, with 23.32 per cent of the votes (after the 24.29 per cent for the Communists). It could get a majority with the help of Liberal Democrats and the Fatherland Russia party. In 2003 United Russia became the largest party, with 38.17 per cent of the votes, while the Communists got only 12.81 per cent. In the election of 2007, United Russia got 64.31per cent of the votes and only 4 parties reached the 7 per cent threshold. In these elections all 450 deputies were elected by proportional vote, whereas earlier half of the deputies were elected from one-person constituencies.

13   Decision of the Constitutional Court No 12-O of 27 June 2000 concerning constitutions of several regions.

14   The president can dissolve the Duma in both situations only after the Duma has voted against the opinion of the president three times.

footnote 13) and that federal legislation granting ownership of natural resources to the Federation does not violate the Constitution.[15] The Constitutional Court has had a significant role in legitimizing the Constitution and giving it a practical meaning.[16]

Local Self-Governance as the Third Level of Public Governance

During the Soviet regime, local administration was only a part of the state apparatus and was coordinated by the Communist Party. Local self-governance was first mentioned in Yeltsin's constitution in articles 81–83, which merely defined local self-governance; self-governance is one of the requirements of the European Charter, ratification of which is required for membership in the Council of Europe.

Even though Russian textbooks on municipal law describe the history of the Russian *zemstvo* (local assembly) before the October Revolution, local self-governance is one of the institutions that has been brought in from outside and represents Western democracy.[17] Many Russian politicians have wondered why the earlier well-functioning local state administration is not enough, which suggests that this new institution has not yet been internalized.

The first federal law on local self-governance did not come into force until 1996, but before that municipalities were given the tasks that they typically have in Western democracies. Municipal infrastructure such as water and sewage, kindergartens and hospitals were often taken care of by large local enterprises. When enterprises were privatized, they had to rid themselves of their duty to run local communities in order to focus on business and increase their competitiveness. The presidential decree of 10 January 1993 prohibited privatization of social and

---

15   The Republic of Karelia and Habarovsk Territory brought the Forest Code of 1996 before the Constitutional Court, claiming that taking forests into federal ownership with a federal law was unconstitutional. Their argument was based on article 72 of the Constitution, which declares that natural resources fall under the joint jurisdiction of the Federation and the region, and article 9, which stipulates that natural resources should be used for the benefit of the people living in the area and that natural resources can be owned privately, by the state or by a municipality. The Court found that the Forest Code did not violate the Constitution, because the regions still receive benefits from forests even though they are owned by the Federation. Secondly, the Court took the view that state ownership of forests has long historical roots in Russia (Decision on 9 January 2008 No3).

16   The Court has been able to give meaning to human rights, also interpreting them according to their wording. It has also handed down decisions that differ from the interpretation of the president. For example, Yeltsin refused to promulgate a law which he had sought to veto, despite the fact that both chambers of the Federal Assembly overruled his veto. The Court held that the president could not refuse to promulgate a federal act when attempts to veto it have been overruled (No 11 P, 6 April 1998).

17   Examples of municipal law textbooks: Ignatov, V.G and Rudoy V.V (2003). Mestnoe samoupravlenie (Feniks: Moskva) and Shumlyankova, N.V. (2002) Munitsipal'noe upravlenie Moskva.

cultural capital along with enterprises. Social and cultural responsibilities were given to municipalities, which were not prepared for the huge increase in their obligations; for instance, there were difficulties in supplying electricity. Many municipalities quite soon started to privatize municipal technology, resulting in pressure to raise the price of these basic services, which had been practically free for Soviet citizens.

The main difficulty in arranging municipal services has been the lack of tax revenues. The companies that had taken care of the local communities around them were no longer able to pay taxes. For a long time taxes were paid largely through an exchange of services. Taxation operates such that municipalities do not get more than half of the tax revenues collected from the local population and enterprises;[18] their expenses for arranging health care, education and municipal technology exceed these revenues. The federal state and the regional state in the middle should subsidize municipalities, but the necessary rules are lacking and subsidizing is arbitrary. The new law on local self-governance, to come into force at the beginning of 2009, should give municipalities more revenues, but it is still too early to know whether the reform will be successful.[19]

Enterprises and the New Business Class

The role of the state in Russia was diminished when enterprises were privatized and a new business class emerged. Privatization had a far- reaching effect on the governance of natural resources in the country. Technically, privatization started already at the end of the 1980s, when workers and managers of state enterprises spontaneously and illegally started to privatize companies for themselves. The government had to react quickly and as soon as B. Yeltsin took office, Gaidar's new reform government introduced a privatization programme, in December 1991 (Krüssman 1998; Radygin 1995). The government aimed at restructuring Russian industry, which was old fashioned, but the managers of state enterprises were such an influential lobby that they managed to persuade the government to give the 'working collectives' an opportunity to become the owners of state enterprises (Radygin 1995; Frye 1997).[20]

Privatization started with presidential decrees, since there was a lot of ideological resistance to privatization in the Supreme Soviet that was elected

---

18   Law on the financial basis of municipal self-governance of 1997. The law will be repealed as soon as the new law on local self-governance of 2006 enters into force.

19   The entering into force of the new law on local self-governance has been postponed because of difficulties in preparing for it in the regions.

20   The original plan would have given workers and managers at most 30 per cent of the shares. Ultimately, the government had to offer alternative programmes and 74 per cent of the enterprises chose the alternative, which gave managers and workers a right to buy 51 per cent of the shares immediately. In many companies the share of insider owners rose to 80 per cent, because insiders managed to buy vouchers and invest in their company.

during Gorbachev's regime. In July 1992, Yeltsin issued a decree on corporatizing state enterprises, giving the enterprises only 60 days to corporatize and submit a privatization plan. In August 1992 he issued a decree on privatization vouchers, which gave every citizen vouchers corresponding to one month's salary (10 000 roubles) to invest in stocks and shares. The vouchers were transferable, which soon created a market for selling and buying them. The whole process was rather disorderly and slipped out of the hands of the government. The result was an insider privatization of the state enterprises to their managers (and workers) and numerous irregularities and abuses. Restructuring was delayed, because the new owners did not have the resources to restructure, and managers even promised their workers that if they supported them in insider privatization, they would keep their jobs and there would be no restructuring (Clarke-Kabalina 1995).

Privatization also created a new class of Russian industrialists, which started to be called 'oligarchs',[21] who managed to acquire valuable assets at the right time. Oligarchs created their own clusters of companies, with a bank supporting them. The law recognized these industrial groups and their creation was encouraged by the government, which was worried about the competitiveness of enterprises that had lost their production chains (Matilainen 2009, forthcoming). The oligarchs supported Yeltsin and cooperation with the weakening political and rising economic power became close and corrupt. When Yeltsin needed support and financing for his campaign for a second presidential term, Vladimir Potanin, the leader of the Onexim Group, suggested to him that the oligarchs could lend money to the state to pay salaries and that the security for the loan would be 29 state enterprises that had not yet been privatized. Every party to the deal knew that the state could not pay back the credit within the one-year period stipulated in the agreement and that the companies would go to the banks (to the oligarchs) (Krüssmann 1998). This was the heyday of the oligarchs' power. After winning the elections, Yeltsin appointed several of them as government ministers.

V. Putin, who was chosen by Yeltsin and supported by the oligarchs (who thought that Putin could be controlled), started a campaign to strengthen the state's power. The richest of the oligarchs, V. Hodorkovskiy, the main owner of the Yukos Oil Company, was charged with tax evasion in 2003 and sentenced to imprisonment in 2005; his oil companies were partly returned to state ownership. Even though Putin taught the oligarchs that they should not interfere in politics, the oligarchs who cooperate with the Kremlin continue to strengthen their empires. Those who have acquired wealth in the oil or aluminium industries have now turned their interest towards other sectors, such as the forest industry, where most companies were privatized and sold to managers and workers. Takeover attempts became intense fights using all kinds of methods to acquire ownership of a company (see Chapter 6).

---

21   Nowadays there are about 300 people who are called oligarchs. Fifty-three of them are on the Forbes list of billionaires.

The badly managed and corrupt privatization process created many problems, such as oligopolistic markets, delayed restructuring and encouraged rough 'grab and run' methods in business. The threat of claims of abuse during the hastily and inadequately regulated privatization compelled many companies to keep a low profile and wait. In the forest industry, although restructuring was delayed, pulp and paper mills started to buy sawmills and logging companies as soon as possible and to lease forests in order to build up production chains that would ensure the availability of raw material. Smaller companies that sprung up as part of privatization have had problems in investing in new technology. The state has encouraged development towards bigger entities, and the industry is now concentrating rapidly.

Today, Russian forest companies can be divided into four groups: those owned by the employees and managers of the mills, those owned by foreign investors, those owned by new Russian forest corporations and those controlled by the oligarchs (Kortelainen and Kotilainen 2003). In Northwest Russia there are two important mills which are still owned mainly by managers and workers: Kondopoga in Karelia and Solombala in Arkhangel'sk. The latter joined the neighbouring Solombala sawmill and wood-processing company in May 2007, forming a new holding group called Solombalales. The companies were one and the same before privatization. The merger is an example of a common process underway in Russia in which larger units are again being formed.

A number of foreign companies have managed to establish operations in Northwest Russia. Examples include International Paper, a North American firm which owns a mill in Svetogorsk in the Leningrad Region, and Neusiedler, an Austrian company (belonging to Mondi Europe, a part of Anglo-American Corporation), which owns the Syktyvkar pulp and paper mill in Komi. Finland- and Sweden-based companies own sawmills, plywood factories and logging companies. Nevertheless, Russia has not been a very attractive target for foreign investors in this sector due to the unpredictability of its rapidly changing societal context and the risky business environment created by the potentially aggressive behaviour of the state and the oligarchs.

In addition to manager-owned companies, Russia has new companies which have bought up old ones and formed new chains. Ilim Pulp Enterprise (now Ilim Group) is the most significant such company; initially it managed to form a group consisting of some of the largest pulp and paper mills in Russia – the Kotlas Mill in Koryazma in the Arkhangel'sk Region and the large Bratsk pulp and paper mill in Siberia. Ilim Pulp announced an initiative to form a holding company with American International Paper (Svetogorsk mill) in October 2007. Another new Russian company, which now has formed a holding company called Investlesprom, started by buying the Segezha cardboard mill in the northern part of Karelia. This new group is controlled by the Bank of Moscow. There have also been intense fights over production facilities. However, the oligarch who has most fiercely tried to enter Northwest Russian pulp and paper industry, Oleg Deripaska, failed in his attempt to take over Ilim Pulp's Kotlas pulp and paper mill (see Chapter 6).

Overall, these developments show an increasing tendency to strengthen the Russian forest industrial sector by bringing in investments and forming holding companies. It seems that concentration under holding companies is going to strengthen Russian companies and has made restructuring possible. It also has made entering the market more difficult, since the existing companies have tried to secure the entire production chain from themselves. The new export duties on roundwood also benefit domestic companies by eliminating competition for raw material and lowering its price.

Civil Society on the Rise

One aim of the Russian transition was to strengthen civil society, which during socialism was under the control of the state and the Communist Party. The Soviet era saw the creation of official associations and societies. Even the labour movement was under state control. Environmental protection, however, had become an important value of Russian scientists, who stood for nature protection against polluting industrialization. This movement started already during Stalin's regime and acted courageously during the years of his dictatorship. With these deep roots in history, environmental protection became an important channel for expressing discontent (Weiner 1999). During Gorbachev's perestroika it was the most visible movement for change, and for the first time, it became a movement of masses. However, the economic and social difficulties which followed the Russian transition made people more interested in supporting their families, and a comparable public interest in environmental issues has not returned since (DeBardeleben – Heuckroth 2001). After the mushrooming of environmental organizations in the early 1990s their number has been on the decline (Tynkkynen 2006). However, this has not meant the demise of such movements, but rather a transformation in their objectives towards building consensus with businesses (see Kotilainen et al. 2008)

Globalization has also affected the development of Russian civil society, with many international organizations entering Russia. Greenpeace and the WWF have established branches and organized campaigns, one of which Maria Tysiachniouk analyzes in her contribution in Chapter 9. Religious and social movements have also entered Russia, both through charity work and by finding new supporters. The spectrum of civil society movements is wide: there are government-supported semi-official NGOs, domestic, spontaneously born NGOs, and branches of international NGOs – sometimes working together and occasionally competing with each other. Chapter 5 in this volume analyzes the activities of NGOs seeking to effect fishery governance in the Murmansk Region.

Suspicion of rivals of the state, especially international movements, by the Putin regime, has resulted in more state control of NGOs. Monetary aid from abroad is nowadays under state control. NGOs are required to report any financial aid from abroad, and supervisory officials are entitled to require any document, send their representative to any meeting and even close down any programme or branch of an international organization if they assume that its work might endanger the constitutional order, morals, health, human rights and interests or the safety of the state.[22] However, NGOs keep on carrying out projects at the grassroots level, as Chapters 5 and 9 of this volume show.

Natural Resources and Interplay Between Institutions and Interest Groups

As the description of the framework of the local level actors shows, they have not been given many tools to influence how natural resources are used. Municipalities are struggling to fulfil their duties with inadequate tax revenues. They may use property taxes to increase revenues and may sell land to (domestic) private buyers. They can also influence the use of land through municipal planning. They are, however, totally dependent on the state and local enterprises, if there are any. It is not difficult to realize that such circumstances encourage corrupt practices both in big cities with a lot of tax revenues and in smaller, remote municipalities with scarce resources.

The federal state has taken all the power to regulate and decide on the use of natural resources through the current federal legislation and with the acceptance of the Federal Constitutional Court. Official institutions thus support highly centralized state governance. The revenues from subsoil resources nowadays go to the federal budget, where the local or regional level may get returns that the Federation agrees to give. The Federation also decides on granting licences for using subsoil resources.[23] With the new Forest Code of 2006, the Federation also took over the power to decide on the revenues from forestry, which, however, can be disbursed to the regions.[24] Municipalities have very few powers indeed when it comes to governing natural resources.

Strengthening and centralizing state governance will not, however, bring back Soviet-style state governance, because in practice enterprises are the key actors in the governance of natural resources. They are the ones that use the resources and interact with both federal and local officials. Enterprises have a lot of pressure to create wealth for both their country and the local community. The state has not

---

22   Law on non-commercial organizations, dated 12 January 1990, as amended 15 May 2008.

23   The law on subsoil resources (oil, gas, coal minerals, and underground water), dated 21 February 1992, originally divided the income among the Federation, the region and the municipality. Based on a study conducted by the Ministry of Natural Resources, which found many errors and abuses in licensing, licensing was brought under federal control with an amendment to the law in 2002. This considerably reduced the income of municipalities with oil resources (Kriukov et al. 2004).

24   The earlier Forest Code of 1997 divided income from the use of forests between the Federation (40 per cent) and the region (60 per cent). Now federal organs hold auctions of forests for lease and of permits to cut ('logging tickets') and the regions get returns from the federal budget.

created an incentive system for enterprises that is as extensive as the one that Western market economies have. Tax incentives and cheap loans are difficult to get, which encourages businesses to network to ensure the availability of raw materials and bank loans. The state encourages enterprises to take on social responsibility, which in practice means that enterprises have to carry the burden of taking care of local communities. The state has not yet managed to build up adequate social services.

Weak property rights still create uncertainty for businesses. The division of powers between the Federation and the regions is still changing even though it now seems that centralization and federal decision power have been stabilized. State property rights are not really expensive for enterprises to use, but typically they have to secure their positions in getting permits for the use of natural resources and use unofficial institutions such as good relations in these operations.

The rules of private property rights are still weak and vague. Russian enterprises and citizens can buy land, but getting an opportunity to do so is dependent on good relationships and other unofficial institutions. Property rights are also weak and vulnerable to claims in courts, as the fate of some oligarchs shows. The state has actively participated in the fight for property rights using methods as dubious as those of the oligarchs, creating more anxiety than trust in the state when it comes to strengthening the legal order.

Using good relations to benefit business is a weak means to secure property rights, especially because Russian officials and politicians often seem to put their personal interests ahead of the interests of the community. Such behaviour is path dependent, since during socialism parasitic use of business income was typical and many local actors do not yet understand the differing rules of a market economy.

Official institutions do not provide tools for the local level to develop. Local self-governance is actually one of the Western institutions that is facing financial difficulties in Western countries as well. Transferring this institution to Russia also transferred the problems, one which are difficult to solve even in the West.

The importance of unofficial institutions is path dependent in a way, because new official procedures do not function properly. The importance of personal relations and good connections with local, regional and federal officials is the way in which business and communities are taken care of. There are, however, new actors, such as NGOs, that also have started to work directly with enterprises without waiting for the state to solve problems first. Forest certification and the social and environmental responsibility connected with it are one strong example of private governance and significant cooperation between private interest groups (For more detail, see Chapter 7).

In this volume we also describe and analyze how local actors have been able to find creative solutions to their problems and survive through difficult times with very little support from official institutions. It remains to be seen how many of these creative solutions prove to be a path to success and how many will remain exceptional solutions adopted during a period of transition. By way of a conclusion, and a starting point for further analysis, we present the figure below showing the changing framework of governance in Russia.

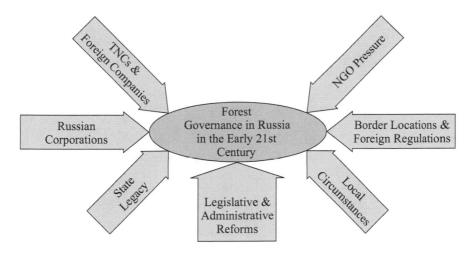

Figure 2.1   Factors Affecting the Forest Governance Regime in Russia

References

Åslund, A. (1995), *How Russia Became a Market Economy* (Washington: Brookings).

Berman, H. (1983), *Law and Revolution. The Formation of the Western Legal Tradition* (Harvard University Press).

Clarke, S. and Kabalina, V. (1995), 'Privatization and the Struggle for Law', pp. 142–158 in Lane (ed.).

Dasgupta, P. and Serageldin I. (2000), *Social Capital* (Washington D.C:World Bank).

DeBardeleben, J. and Heuckroth, K. (2001), 'Public Attitudes and Ecological Modernization in Russia', pp. 49–76 in Massa and Tynkkynen (eds).

Frye, T. (1997), 'Russian Privatization on the Limits of Credible Commitment', pp. 84–108 in Wilmer (ed.).

Furman, D. (2007), 'The Origins and Elements of Imitation Democracies. Political development in the post soviet space', *Osteuropa (special issue). The Europe beyond Europe. Outer Borders, Inner Limits. Osteuropa* 205–243.

Granovetter, M. (1985), 'Economic Action and Social Structure: the Problem of Embeddedness', *American Journal of Sociology* 91:3, 481–510.

Hausner, J., Jessop, B. and Nielsen, K. (eds) (1995), *Strategic Choice and Path-Dependency in Post Socialism. Institutional Dynamics in the Transformation Process* (London: Edward Elgar Publishing).

Hooghe, L. and Marks, G. (2003), 'Unraveling the Central State but How? Types of Multi-Level Governance', *American Political Science Review* 58:3, 189–193.

Hosking, G. (1985), *A History of the Soviet Union* (Glasgow: Fontana Press).

Kaabolian, L. (1998), 'The New Public Management: Challenging the Boundaries of the management versus administration debate', *Public Administration Review* 58:3, 189–193.

Van Kersbergen, K. and van Waarden, F. (2004), "Governance' as a bridge between disciplines: Cross-disciplinary inspiration regarding shifts in governance and problems of governability, accountability and legitimacy', *European Journal of Political Research* 43, 143–171.

Kosonen, R. (2002), *Governance, the Local Regulation Process, and Enterprise Adaptation in Post-Socialism: the Case of Vyborg* (Helsinki: Helsinki School of Economics A 199).

Kortelainen, J. and Kotilainen, J. (2003), 'Ownership changes and local Development in the Russian pulp and paper industry', *Eurasian Geography and Economics* 44:5, 384–402.

Kotilainen, J., Tysiachniouk, M., Kuliasova A., Kuliasov, I and Pchelkina, S. (2008), 'The potential for ecological modernisation in russia: scenarios from the forest industry', *Environmental Politics* 17:1, 58–77.

Kriukov, V., Seliverstov, V. and Tokarav, A. (2004), 'Federalism and Regional Policy in Russia: Problems of Socio-Economic Development of Resource Territories and Subsoil Use', pp. 96–127 in Peter H. Solomon, Jr. (ed.).

Kryshtanovskaya, O.V. (2004), *Anatomiya Rossiskoi elity* (Moscow).

Krüssman, T. (1998), *Privatisierung und Umstrukturierung in Russland. Zur Rolle des Rechts als Instrument Struktureller Wirtschaftsreform in Übergang zur Marktwirtschaft* (Berlin Spitzverlag und Vienna: Verlag Österreich).

Lane, D. (1995), *Russia in Transition* (London: Longman).

Lonkila, M. (1999), *Social Networks in Post-Soviet Russia: Continuity and Change in the Everyday Life of St. Petersburg Teachers* (Helsinki: Kikimora Publications).

Långström, T. (2003), *Transformation in Russia and International Law* (Leiden: M. Nijhoff Publishers).

Massa, I. and Tynkkynen V-P. (2001), *The Struggle for Russian Environmental Policy* (Helsinki: Kikimora Publications B17).

Matilainen, A-M. (2009), 'Holding companies in the Russian Forest Industry Sector: From Timber Harvesting to End Products', unpublished contribution forthcoming in the book of the 'Russia in Flux' Programme of the Academy of Finland at Brill Publishers.

McFaul, M. and Perlmutter, T. (eds) (1995), *Privatization, Conversion, and Enterprise Reform in Russia* (Colorado, USA: Westview Press).

Nielsen, K., Jessop, B. and Hausner, J. (1995), 'Institutional Change in Post-Socialism', pp. 3–46 in Hausner et al (eds).

North, D.C. (1990), *Institutions, Institutional Change and Economic Performance: A New Economic History* (Cambridge: Cambridge University Press).

North, D.C. (2005), *Understanding the Process of Economic Change* (Princeton and Oxford: Princeton University Press).

Nozick, R. (1974), *Anarchy, State and Utopia* (New York: Basic Books).

Nystén-Haarala, S. (2001), *Russian Law in Transition. Law and Institutional Change* (Helsinki: Kikimora Publications B:21).

Oleinik, A. (2001), 'Biznes po panyatiyam: ob institutsional'noj modeli Rossiiskogo kapitalizma', *Voprosy ekonomiki* 5, 4–25.

Ostrom, E. (1990), *Governing the Commons: The Evolution of Institutions For Collective Action* (Cambridge: Cambridge University Press).

Radaev, V. (2002), 'Rossiiskii biznes: Na puti k legalizatsii?', *Voprosy ekonomiki* 1, 68–87.

Radygin, A. (1995), 'The Russian Model of Mass Privatization: Governmental Policy and First Results', in McFaul, M. and Perlmutter, T. (eds).

Rhodes, R.A.W. (1997), *Understanding Governance. Policy, Networks, Governance, Reflexivity and Accountability* (Buckingham: Open University Press).

Rose, R. (2000), 'Getting Things Done in an Antimodern Society: Social Capital Networks in Russia', pp. 147–171 in Dasgupta and Seragelden (eds).

Rosenau, J.N. and Czempiel, E.O. (eds) (1992), *Governance without Government: Order and Change in World Politics* (Cambridge: Cambridge University Press).

Sachs, J. (1993), *Poland's Jump to the Market Economy* (Cambridge, Mass.: MIT Press).

Shulzhenko, Yu., L. (1995), *Konstitutsionnyi kontrol v Rossii* (Moscow: RAS, Institut gosudarstva i prava).

Solomon, P.H. Jr. (ed.) (2004), *The Dynamics of 'Real Federalism'. Law Economic Development, and Indigenous Communities in Russia and Canada* (Toronto: University of Toronto, Centre for Russian and East European Studies).

Stiglitz, J. (1993). *The Wither Reform? Ten Years of the Transition* (Washington DC: World Bank Annual Conference on Development Economics, 28–30 April).

Stiglitz, J. (2002), *Globalization and its Discontents* (Great Britain: Allan Lane, Penguin Books).

Sutela, P. (2003), *The Russian Market Economy* (Helsinki: Kikimora Publications B:31).

Tynkkynen, Nina (2006), 'Action frames of environmental organisation in Pos-Soviet St. Petersburg', *Environmental Politics* 15:5, 639–649.

Weiner, D. R. (1999), *A Little Corner of Freedom. Russian Nature Protection from Stalin to Gorbachev* (Berkley/Los Angeles/London: University of California Press).

Williamson, O.E. (1985), *Economic Institutions of Capitalism* (New York: The Free Press).

Williamson, O.E. (2000), 'The New Institutional Economics', *Journal of Economic Literature* 38:3, 595–613.

Wilmer, D.L. (ed.) (1997), *The Political Economy of Property Rights* (Cambridge: Cambridge University Press).

Chapter 3

The Task of Macroeconomic Policy in Generating Trust in Russia's Development

Stefan Walter

Introduction

The overall performance of the Russian economy has been striking and is now in its eighth year of expansion. Russian Gross Domestic Product (GDP) grew by about seven per cent per annum during the period 1999–2006 (BOFIT 2007). This is much more than could have conceivably been expected after the strong contraction of the Russian economy in the 1990s and the 1998 crisis. The economy was driven particularly by growth in industrial production and export volumes. Natural resource extraction especially has contributed to this trend – around 70 per cent of the industrial growth between 2001 and 2004, with the oil sector accounting for 45 per cent. Oil is also the most important export commodity and thus mainly responsible for the growth in export volumes (Ahrend 2006).

Russian exports will remain large in the foreseeable future, as the main export commodity is oil. The favourable oil market price has resulted in a strong surplus in Russia's current account; a large surplus has been the norm since 2000. Governmental spending was cut around the same time. A stabilization fund has been put into place that receives a large part of the surplus from windfall revenues and ought to assist in coping with economic shocks. At the end of 2005, this fund was calculated at about 50 billion US dollars. The recent success of the Russian economy is estimated to be 50 per cent due to the development in the price of oil and 50 per cent to responsible economic policy, price competitiveness generated by the rouble depreciation after the 1998 currency crisis, institutional changes and a general recovery from the economic trough of the 1990s (Sutela 2005).

In contrast to other industrial sectors, the forest industry is less dependent on oil and gas. The industry is organized as a cluster that includes harvesting, mechanical wood processing and the pulp-and-paper industries. The forest cluster as a whole is one of the largest industries in Northwest Russia, producing approximately 15 per cent of the total industrial output in recent years. Moreover, the cluster is more developed than in other Russian regions; one advantage is its proximity to the European market (Dudarev et al. 2004).

The forest cluster has somewhat suffered from a lack of investment. Consequently, its range of high value-added products is modest compared with that of the world leaders in forestry. There is great potential for growth in the Russian forest cluster, but this is very much dependent on a growing domestic economy, which ought to provide the financial capabilities for an upgrade of the product range. In any case, so far the forest cluster has been dependent predominantly on exports made up mostly of low value-added products, such as roundwood, sawn timber, and paperboard, but it is more evenly balanced in Northwest Russia than in other Russian regions. However, the domestic market demand in Russia on the whole has been decreasing throughout the last decade, which had a marked effect on sales in general, not only in the forest cluster. Nevertheless, there has been some growth and development of prices and product segmentation in the forest cluster, particularly in the St. Petersburg region, which is a major centre of consumption in Northwest Russia (Dudarev et al. 2004).

The competitiveness of the Russian forest cluster has been rarely based on advanced, value-added products. Since the beginning of the 1990s it has been based on the utilization of basic factors, most notably the extensive forest resources, cheap labour, energy and transport. However, these costs are bound to rise in the not too distant future (Dudarev et al. 2002). The success of Russian exports in recent years is going to weaken the price competitiveness of Russian forestry products. The strong dependence on exports makes products more expensive and less desirable; although, if new technologies are applied – and new technologies are overwhelmingly of foreign origin – a higher exchange rate can provide an advantage for the forest cluster, provided that the cluster produces more for the domestic market (Holopainen et al. 2006). One incentive to make wood available to the domestic market is the recent policy of the Russian federal government to tax the export of roundwood – an attempt to protect the Russian domestic industry from the appreciating exchange rate.

On the whole, production facilities, infrastructure, and training of workforce date mostly from Soviet times. However, these important input factors are largely depleted due to insufficient investments. Requirements for a more solid foundation for the cluster include upgrading of equipment and production facilities, construction of new transport infrastructure and/or maintenance of the existing infrastructure, reforestation measures and other environmental services, and the training of specialists and labour to work with contemporary production requirements .(Dudarev et al. 2004) Hence, one of the most urgent issues for the further development of the forest cluster is the improvement of the investment and business climate in order to increase the level of investment, including foreign direct investment. Unfavourable investment conditions prevent an inflow of capital. At present, investment focuses on urgent activities necessary for maintaining the existing operations in the cluster. Furthermore, regional development in the form of concentration and specialization is a major trend anticipated in the forest cluster of Northwest Russia (Dudarev et al. 2002). Sooner or later prices for input factors will rise and will increasingly lead to a loss of competitiveness of enterprises in the forest cluster.

Yet there will remain differences in the rate of development between regions. It is envisaged that production efficiency between forest clusters in Russia and other countries will probably decrease in the next ten years. Therefore, corrective measures are required to overcome the problem of rising prices. One such measure is the fostering of innovation in the cluster. Currently the forest cluster lacks the prerequisites for modernization, due to the cluster's present high profitability being based on the exploitation of its competitive advantages, i.e. cheap wood, energy and labour. In the long term this will have an adverse effect, but this is at present economically not yet relevant enough. However, the advantages are likely to diminish and companies will have to undertake measures to remain on the market, where modernization is the most important consideration. (Dudarev et al. 2002).

The situation in the forest cluster reflects the situation overall in Russian industry. On the whole, exports, albeit successful and rising, appear rather one-sided. The prevailing opinion is that this is not due to an overvalued currency; rather, the currency is still sufficiently undervalued to continue to drive exports. Lack of competitiveness is considered to prevent many sectors from exporting more than they do. (Sutela 2005). This can even be observed in the most successful sector, the oil industry. The largest contribution within this sector to economic growth and exports was made by the private oil enterprises, which are mostly controlled by the financial sector. These are the ones that have received the much-needed investments and experienced a change towards an efficient business model in order to increase the level of competitiveness. On the other hand, the state oil sector has been rather unable to make a sizable contribution to the overall success of the Russian oil industry (Ahrend 2006).

In any case, economic growth in Russia will be lower in the near future than it has been in previous years. Increasing capacity utilization cannot continue forever. Policies should focus on motivating investments so that economic growth can change from being based on increasing capacity utilization to being based on investments. Even though, as noted, productivity can still be enhanced for a long time, an increased propensity for imports and a consequent appreciation of the rouble suggest that Russia cannot remain a cheap output country for long (Sutela 2005). Good macro-economic policy is without doubt crucial here (Ahrend 2006): it has to create the conditions to provide necessary financial capital, both for the maintenance of infrastructure and production facilities and for the further improvement and development of new products and production processes. However, financial capital is only one factor among those needed to create this favourable investment climate.

In this chapter, I would like to focus particularly on the notion of trust. Trust is basically the confidence in one's expectations and it is a basic requirement for social life. Trust cannot be assumed to exist; i.e. it cannot be assumed that people inherently bestow it. Rather, trust has to be built up and maintained. Secured expectations produce social order through which governance is possible (Jalava 2003). Without sufficient trust in institutions, successful governance, for example, aiming at sustainable development, is jeopardized. Here, I aim to demonstrate that macroeconomic policy has important tasks in generating the trust required to allow healthy economic development in Russia. Therefore, I will delineate broad developments in Russian macroeconomic policy to illustrate its trust-generating potential. First, however, I will start with a conceptual definition of trust and how it relates to the economy and to economic policy.

System Theoretical Concept of Trust

Trust can be understood as a mechanism that serves to reduce social complexity. Trust allows a higher complexity of the human experience; more possibilities of acting and decision-making can be realized in order to allow any increases in social complexity (Luhmann 2000). Complexity is a condition of systems; systems build up complexity in the course of their evolution. Trust is located in the relations between people; it is not a psychological state of isolated individuals. Accordingly, trust must be seen as a property of collective units – of systems. Social relations depend on trust. Trust can be viewed as a prerequisite for the proper functioning of society (Lewis & Weigert 1985). The alternative to trust in social relations is chaos. And no one trusts chaos (Luhmann 2000: 47).

Increases in social complexity and reductions of that complexity through certain mechanisms go hand in hand. Simultaneous increases and decreases can be seen as a necessity of social structure, of the structure of human social behaviour. It allows taking into account multiple decision-making contingencies of the self and the other in an interactive situation. In order to increase the probability of successful interaction, this *social dimension* of trust requires a common communication standard. The symbolically generalized media of communication provide this standard. In the case of the economy, money is the medium of communication (Luhmann 2000).

It is the function of the symbolically generalized media of communication to motivate the *autopoiesis* – the reproduction – of operations in systems. Depending on the success of the motivation to reproduce the system, it is possible that the medium is used too much or too little. With respect to the economic system this can be easily illustrated by referring to the terms inflation (too much usage) and deflation (too little usage). The source of the problem in this case is not a lack of coverage for an underused or overused medium – money – through real goods or gold. Rather there is a lack of trust in relation to the possibility of continuing the use of the medium, that is, the continuation of the circulation of the medium. Inflation occurs when continued communication – payments or investments – require more trust than the medium can produce. In this case the medium (money) becomes devalued (expressed as a price increase). In turn, deflation occurs when the communication leaves opportunities to produce trust untouched. In that case the medium is circulated less, with the future opportunities for using money as medium for payments decreasing. The verge of either inflation or deflation is reached when the conditions for the continuation of autopoiesis in the system become so strict that they do not permit further autopoiesis. These conditions are called hyperinflation and hyperdeflation, respectively. They reflect the situation before symbolically generalized media of communication emerged and they reflect the improbability of successful cooperation. The difference between now and then is that modern society is not prepared structurally to deal with the case of the improbability of autopoiesis. This can have serious repercussions for trust in other systems, for instance, the political system (Luhmann 1997).

Thus, in conditions where trust is eroding, there is a serious threat that society could disintegrate. There is, therefore, a risk involved – a risk generated by the lack of trust. Risk would not exist if there were a functional alternative to trust. The risk becomes clear when thinking about planning – a task for the political system of society. Planning would not be possible if the planner had to take into account all possible contingent futures (Lewis & Weigert 1985). This would entail dealing with an infinite complexity, which is something that a system attempts to reduce.

Rational prediction is a possibility to overcome the problem of high contingency. Another solution is to incorporate trust into the planning procedure. Where rational prediction alone would fail, trust becomes in fact vital for planning and decision-making. (Lewis & Weigert 1985). Thus, this confirms the earlier statement that trust reduces social complexity.

Of course, an erosion of trust, or emergence of distrust, might occur, but society is not conceivable without a fundamental basis of trust. In fact, distrust can be helpful in some cases, for example in politics, where a democracy is based on a 'healthy' distrust and change of the parties in power. But even democracy is not possible without a basic trust that politics can actually fulfil its tasks when allocating power to an authority like a government. A betrayal of trust acts as a complete blow to the foundation of a social relationship (Lewis and Weigert 1985). Trust works as a foundation for society upon which social relations, planning, justice, etc. are crucially dependent. Society thus has an interest in maintaining or strengthening trust in systems and institutions – also symbolically generalized media of communication are such institutions.

Trust also affects time. This can be easily interpreted as trust in the future. Trust here is concerned with the future of a certain present; it is the attempt to realize that future and make it present. What appears complicated is planning. Planning is the prediction of a possible future and is highly problematic in a societal setting with rising complexity. Increasing complexity necessitates deferral, such as a deferral of needs; time allows an ordering of decisions and events into a succession because, with higher complexity, fewer possibilities can be realized in a given present. More complexity demands more certainty; trust produces that certainty. It stabilizes the present, which is at the same time associated with a certain past and future, describing what can be almost called an era. This *temporal dimension* of trust, however, is under threat due to the emergence of a social orientation based on a rational-scientific-technological paradigm. This paradigm increasingly dominates social life at the expense of the present. The paradigm produces a general acceleration of social events and a simultaneous strong demand for trust to adhere to a certain present. An attempt to deal with this problem collectively has been the planning and organization of time as expressed through certain ideological orientations, such as socialism (Luhmann 2000).

Modern society, which is differentiated into subsystems fulfilling different functions has a high capacity to solve problems and can see the world in more complex terms. Such a capacity, however, is only possible if certain selections (or contingencies) can be handled in advance; thus, certain decisions should already been made for the decision-maker. High complexity presumes that a large number of choices is possible – too many to be left to the individual. Symbolic communication media, such as money, are evolutionarily successful mechanisms that create expectations and motivations in interactive situations. Such media do not have to fall back on interpersonal trust; in fact, here, personal relationships do not matter. This makes the media increase the probability of successful communication (Luhmann 2000).

Thus, mechanisms like money permit living in a future of high complexity. Money is transferable freedom against a limited choice of goods. Freedom means here that a selection is open to individual decision-making. The social dimension and the temporal dimension of trust converge here: a deferral of present needs for the sake of future consumption (time) is coupled with *not* having to know the multiple behavioural contingencies of the other (social), as long as trust in money exists. The trust in the medium is the trust that the "system works". Thus, trust in the system represents a transformation of personal trust into system trust. One consequence of this is faster learning and information processing; less information is needed for decision-making if money is available (Luhmann 2000).

However, investments have to be evaluated according to their potential consequences. A basic principle is that investments mean a loss of liquidity and, hence, freedom of choice. A possibility to overcome this principle (and to create a 'paradox') is to gain and to lose liquidity simultaneously when investing money and receiving interest on it. This can only be achieved through an increase in the volume of money circulating in the economy. This is followed by an increase in complexity due to the rising volume of money and a resulting threat to the maintenance of trust (Luhmann 2000). Lewis and Weigert (1985) are of the same opinion when they write that any interference – and increase of the volume of money due to interest rates is interference – is bad for the trust in money. The uncertainty associated with the decision to invest or not significantly influences whether trust can be maintained. The uncertainty has two possible effects: it can lead to a lack of investments (deflation) or to a flight into real assets (inflation). This dual effect makes precise control of investments difficult if not impossible. Individual investment decisions are, hence, very much dependent on trust in the economy's currency (Luhmann 2000). From this it follows that trust needs to be maintained through policy – economic policy.

Economic Policy Considerations

Traditionally, economic theory has centred on production. Thus, production and, in some flavours of the theory, labour, as a production factor, have been the focus

(Luhmann 1988). Consequently, one concern of economic analysis has been the expansion of production and economic growth in material terms and, as a result, also of economic policy considerations. For example, economic policy might focus on keeping the exchange rate at a level where it creates favourable conditions for domestic exporters at the expense of currency stability by accepting some inflation. The theory of social systems, in turn, focuses on money as a communication medium. Money is at the centre of the economy, hence the importance of trust in money as maintaining the economy's function, which is to satisfy needs over long time spans. Thus, the stability of money – in contrast to increasing production – should be at the centre of economic policy. In other words, the purpose of economic policy is to maintain trust.

As soon as money emerges in evolution, solvency and insolvency, or liquidity and illiquidity, are generated. These create a double cycle of passing on solvency and insolvency in different directions. This double cycle is managed by banks. Banks have been created due to the interest rate problem; they can solve the problem of how to convert illiquidity into liquidity. And in doing this they possess an exclusive privilege; this has prompted some to call banks parasites since they sell their own illiquidity to others, their customers, at a profit by borrowing money from the central bank. The central bank, in turn, must regulate the management of liquidity and illiquidity, including the extent to which banks can borrow, lend, and charge for money. In other words, the central bank must regulate the extent to which the conversion of illiquidity into liquidity can be a profitable activity, without being itself bound to profitability (Luhmann 1988).

Ecological issues, with which natural resource governance is concerned, make governance of society or its parts, such as the economy, indispensable. Governance, in system theoretical terms called steering, is first and foremost *self-steering*, since social systems are self-organizing and organizationally autonomous. There is, however, a tradition to call for politics to affect societal conditions and to establish social policy. But this conflicts fundamentally with the self-organizing principles and the resulting limited capacity of politics to govern other parts of society than itself. If aiming to influence a system outside politics, politics must take into account the other system's distinctive features, that is, how that system differentiates itself from its environment.

In the case of the economy, the difference is, as noted, the difference between liquidity and illiquidity. Any effort to establish a political programme that ought to affect the economy must be based on this difference. Governance in this case means to reduce that particular difference. Reducing the difference is essentially what activates a system. In the economy, all economic activity aims to balance liquidity with illiquidity. This principle has to be utilized by politics as well. It is in fact the only way to govern. Thus, a difference expressed in a monetary terms and established in a political programme can affect how profitable an investment is. For example, an ecological programme might introduce a tax on a raw material, increasing the price and reducing the profit. Consuming more will be a less profitable activity. However, it is not possible to reduce the difference to zero. The difference will always be there; it can only be minimized In the case of complete reduction, the logic according to which the economic system distinguishes itself from its environment would be removed, leading to destruction of the system (Luhmann 1988).

Any political programme aiming at decreasing the economy's ecological impact would have to address the speed of resource consumption. Scale matters in resource management, as Daly (1992) writes. The idea of scale is based on the understanding that there is a temporal dimension associated with natural resources. This dimension also incorporates the variation that is so common in natural phenomena, as well as cycles. The time inherent in cycles constrains the possible consumption by humans. Depending on how large these regeneration cycles are, we commonly differentiate them by defining resources as renewable or non-renewable (on the scale of a human lifetime) (Jordan and Fortin 2002). Resource consumption is reflected in harvesting cycles and these are tied to the economic logic of time compression. It seems that the economy strives constantly to gain time in order to reduce the time lag between investment and the restoration of solvency. With respect to the forestry industry, one potential consequence is the acceleration of wood harvesting.

Money is a reflexive medium, which means that it can be applied to itself – it is possible to buy money with money (Luhmann 1970). The price of money in relation to monetary inflation (or deflation), reflected in the central bank's policies on interest rates, determines the amortization rate of investments. Other things being equal, a high price for money (again, relative to the inflation rate) requires a longer amortization period – and thus allows less consumption within a given time period – than a lower price for money. Policies that manage money, therefore, have a function in potentially affecting the speed of natural resource harvesting (Walter, 2008).

Politics can still make wrong assumptions when designing such policies. Especially in the context of creating development policies, politics might make the assumption that it is sufficient to provide a high profit to attract investors. However, such a development programme can easily fail, providing evidence that it is not the highest profit that attracts investment (see Walter, 2008, for evidence). Enterprises and households aim primarily at reducing their risk, which is also the purpose of a modern economy.

Function of Macroeconomic Policy

Thus, in sum the primary function of policies to manage money lies in maintaining the motivation to use money as a means of value exchange in the economy. In other words, policies ought to prevent massive changes in the value of money, such as deflation or inflation, to make sure that the economy continues to operate through the usage of money (Luhmann 1984). Ensuring the continuing usage of money is based on the understanding that the flow of money in the economy is of a cyclic nature (e.g. Luhmann 1986; Woodruff 2005). Using money for a payment requires that a payment has occurred in the past, since it is only possible to spend money if one has earned money through someone else's money transfer.

The motivation of the utilization of money has a twofold effect. First, keeping the value of money stable generates trust in the medium and the performance of the economy. Trust is based on expectations; these make the world less chaotic and more predictable. This enables social order, through which steering (governance) is possible (Jalava 2003). Trust, however, is difficult to achieve if institutions have a low credibility. A functioning economy, for example, is fundamental to restore credibility. This is illustrated by the savings ratio of Russians. The combined assets of Russian banks were around 35 per cent of Russia's Gross Domestic Product (GDP) in early 1998. In comparison to other countries, even those that are still considered emerging economies, this is a rather low ratio. A large part of the small assets that banks had was invested to finance the deficit of the public economy. In turn, a comparatively small part of those assets were lent to private enterprises. That means that not enough of the small amount of available capital was used to finance the economy (Komulainen 1999). Furthermore, the use of monetary surrogates on a surprisingly large scale during the 1990s prompted some to call the Russian forest economy a virtual economy. Approximately 50 to 70 per cent of Russian industrial production in August 1998 was exchanged through barter (Woodruff 2005); this is largely an issue of the past and barter, as well as the virtual elements of the forest economy, has gradually disappeared with the transition (Mashkina 2006).

The second effect of maintaining monetary stability is the provision of a structural foundation for resource governance. This is directly connected to the generation of trust; i.e. only through trust is a certain degree of steering possible. This shows that governance is a path-dependent activity. For example, the use of a stumpage fee to artificially increase the price of wood, the rationale being that a higher price will cause a commodity to be used more efficiently, naturally assumes that money is actually used to obtain the right to the wood. If money is not used, as in the cases of monetary surrogates, a stumpage fee would be essentially useless. Governance, therefore, assumes that systems (as path patterns of social behaviour) exist and continue to exist (e.g. through the continued use of money).

Accordingly, governance must be understood as a mutual activity. A societal achievement is not considered to be the result of purely political operations, but, rather, dependent on the interaction and combination of other sectors (systems) of society. To produce successful outcomes, governance has to count on the contribution of all parts of society, where these represent in fact all the resources that are available in society to solve a problem. One could now formulate the task of governance as being the facilitation of the interaction of different societal spheres. One consequence of this view is that politics is not to be considered a superior system in society and that a central government is not supreme. Society is in fact without a centre (Rhodes 1996; Kooiman 1993).

When aiming to integrate the variety of systems into a common endeavour, one has to accept that there are limits on the extent to which society can adapt. These system rationalities have to be accepted and cannot be bypassed when aiming for a successful governance effort. Thus, in line with the idea of a need for system continuity (path-dependence) in the concept of governance, one must be aware that those functional systems with their peculiar rationalities represent institutional developments on whose successful operation governance is dependent. If for some reason the operations cannot be carried out – or can only be carried out under difficult circumstances, such as when power relationships are unclear due to legal failure or legal inaccuracies or when barter is used instead of money for trading – the success of governance will be greatly jeopardized. For example, resource consumption in the informal economy falls either outside the government's monitoring ability or is more difficult to control. It must, therefore, be in the greatest interest of any government to assist the support of formal institutions by the public.

Macroeconomic Policy in Russia

Macroeconomic policy as described in this chapter includes fiscal and monetary policies. Generally, such an analysis might also include the relations of the various institutions involved in managing monetary value and in planning budgets, the different levels of government (federal and regional), central bank, and other stakeholders, including the financial system involving commercial banks. Macroeconomic policy has several important functions with regard to the management of natural resources such as forests.

During the Soviet era, monetary policy had two main roles: One was to ensure the fulfilment of the economic plan; the other was to control the purchasing power of Soviet households. In a centrally planned economy the plan is, of course, the central institution serving as the guideline for production numbers and prices and, at the same time, as the basis for the allocation of credit to producers in order to achieve the production goals. Supplying enterprises were paid through bank transfer; money, in turn, was only used by enterprises to pay wages and salaries to workers and employees. The second role existed to avoid queues and shortages of supplies. To achieve this, monetary policy targeted the amount of cash in circulation. A cash plan existed in which the head organization in the Soviet banking system, Gosbank, established how much cash would be allocated to enterprises so that they could pay their personnel (Baliño 1998).

Gosbank fulfilled the roles which are split between the central bank and commercial banks in states with market economies: to issue the money, to clear transfers between enterprises, to transfer credit, and to formulate the cash plan. Gosbank controlled other banks in the Soviet Union, including several specialized banks that financed different industrial sectors, as well as the Savings Bank, where households could deposit some of their cash savings. Furthermore, an official exchange rate plan existed in which the rate was administratively set so as to ensure that domestic prices would be equivalent to international market prices.

This was achieved through subsidies and taxes (Baliño 1998). When the Soviet Union dissolved at the end of 1991, Russia had to newly create or adapt its monetary institutions to new political and economic realities.

Fiscal Policy

Russian fiscal policy since the early 1990s has been very unstable and unpredictable, partly due to a lack of consensus within the state apparatus as to the role of government in the economy and partly due to a sheer lack of alternatives to compensate for the high fiscal deficits other than using central bank credits (Sutela 2003; Baliño et al. 1997). For a long time, the Russian federal government followed an expansionary fiscal policy course, partly to finance the budget deficit and partly to keep the rouble within the currency band of an exchange rate programme; however, the rate of central bank credit to the government accelerated and slowed down erratically (Baliño et al. 1997). This monetary expansion, of course, fuelled inflation – not only when the actual borrowing occurred, but also thereafter through a delayed impact (Orlowski 1997). The growth of money was thus caused to a significant extent by a need to meet the budget's obligations, not so much by a demanding economy. The variation in fiscal policy illustrates well the problems in overcoming the legacy of central planning where budget and credit financing were not distinguished. The problems reflect the use of macroeconomic policy by authorities as an instrument of social welfare provision, which was reflected in a lack of fiscal regulation (Granville and Mallick 2006). Given the function of financial policies to include the maintenance of an integrated economy by motivating the use of money, it is possible to conclude what the consequences of fuelling inflation were for the industry. The low levels of investment in the forest sector and the extensive use of barter during the 1990s indicated the lack of a valuable currency. Thus, much of the sector had little choice but to 'invest' using relational capital, thus missing out on real investments in physical capital that could have strengthened its competitive position in the market for forest products.

This fiscal behaviour produced great difficulties for the monetary authorities to pursue their policy of controlling the mass growth of money through reserve requirements. By receiving large amounts of money through free credits that were covered by the budget (in 1998 and 1999 these equalled nine and three point three five per cent, respectively, of the Russian GDP) the economy had few if any additional requirements for borrowing money from banks. The banks, in turn, had no need to participate in credit auctions, as they possessed sufficient liquidity (Aleksashenko 2000). For example, at the beginning of 2004 real interest rates were still negative and reflected excess market liquidity (BOFIT 2004). The fiscal behaviour of the government resulted in the Central Bank having fewer possibilities to govern the supply of money to the economy, since monetary aggregates represent the instrument of choice for the Russian Central Bank. The critical point here is that at least until the mid-1990s there was no real independent monetary policy that the Central Bank could pursue as long as its behaviour was programmed to adjust to the government's budget needs. This constellation had to be considered unsustainable (Aleksashenko 2000). It also illustrates that the federal government, irrespective of the Central Bank's goals, had no real interest in controlling inflation. The focus of the government was on sustaining enterprises and their production, which occurred at the expense of maintaining trust and strengthening currency stabilization.

The mid-1990s saw an increasing consensus among previously conflicting interest groups regarding inflation policy. Macroeconomic policy was developed with the intention of lowering inflation by decreasing the rate of mass money growth in addition to cutting the fiscal deficit, tightening monetary policy and further liberalizing trade and prices. One reason why there was suddenly a much stronger consensus on how to proceed further was that the policy-makers had learned about the real costs of unhealthily high inflation rates. For example, inflation led to increasing dollarization and use of barter as alternatives to the rouble. The use of rouble alternatives decreased the tax base in the economy; no government could be interested in this. Attempts were made to finance the budget deficits by other means, for example, treasury bills; these were initially short-term but long-term bills were planned. The federal government made the mistake of continuing to borrow while the yields and attractiveness of the treasury bills were high; it did not attempt to continue balancing the budget. This was not interfered with, as it was believed that although the debt-to-GDP ratio rose dangerously, the ratio would decrease to moderate levels in the time to come, given that the economy had already showed signs of a small recovery. The crisis in Asia then affected the Russian financial system in a catastrophic way. It also highlighted the fact that many reforms had not been undertaken yet that could have helped to improve the situation, including tax reform, bankruptcy law, and land property issues. Lower energy prices and rising rouble interests followed the crisis in Asia, as the financial market became nervous about Russia's capability to deal with the issues. Towards the height of the rouble crisis, debt servicing took almost all tax revenues. This in turn made credit financing from the central bank a necessity. Because the structural reforms mentioned were too slow to revitalize the economy, debt became excessive. In August 1998 Russia had to declare itself in default. This was followed by a devaluation of the rouble by 70 per cent, making import prices four to five times higher than before the devaluation (Sutela 2003).

Luckily the anti-inflationary consensus in Russia held and there was no attempt to raise wages to counteract the price increases. After 2000 the Russian government introduced fiscal reforms, which have produced sizable surpluses based on a steady rise in oil prices. Furthermore, favourable trade conditions and growth contributed to a balanced budget. For example, revenues, including value-added tax, increased moderately in 1998, while expenditures stayed fairly equal and debt service expenditures even decreased; the developments combined to result in a decreasing deficit. Fiscal institutions were reformed, with this including an overhaul of the tax system, the introduction of a natural resource extraction tax, the abolition of turnover tax for enterprises, and a new budget code that demanded more fiscal responsibility and restricted government spending and borrowing (Ahrend 2004).

Monetary Policy

The Rouble Zone

After the Soviet Union was dissolved, it was initially decided to keep up a unified monetary system, the rouble zone, for most of the post-Soviet republics. Gosbank disappeared and central banks were installed in the member countries of the rouble zone as the institutions responsible for monetary policy. In this arrangement, the Central Bank of Russia became the sole issuer of cash. However, all central banks of the participating countries could grant credit. This led to a situation of one monetary zone with multiple money-creating central banks, resulting in incentives for the smaller central banks to expand credit to promote economic development in their respective jurisdictions (Baliño 1998; Woodruff 2000).

The outcome of this regime was monetary inflation that spilled over the whole rouble zone, and developments quickly moved towards a centralized solution. Initially, the Central Bank of Russia confined itself to controlling the internal credit flow between all the central banks. By mid-1993, however, the massive problems in the rouble zone, also exacerbated through the financing of fiscal deficits, prompted Russia to introduce the Russian rouble and demonetize the pre-1993 roubles of the unified monetary system. This marked the emergence of an independent Russian monetary policy (Baliño 1998). Until then the Central Bank was not realistically capable of dealing with inflation; this shortcoming was also due to the lack of a strict hierarchy in the economy with respect to the control of money – a hierarchy which is vital when aiming to control money flows between one centre and business banks, on the one hand, and banks and enterprises and households, on the other. This centralized management of money and the decentralized decision-making on investments is in fact the great advantage of money in the modern economy and one reason for its evolutionary success.

The Fixed Exchange Rate Programme

In 1995 Russia adopted an exchange rate policy after the three-year period of high inflation that followed the break-up of the Soviet Union and the establishment of the unified monetary area. A fixed exchange rate programme, used by many states with emerging markets, is designed to control inflation by linking the domestic currency to an authoritative international currency, which is in most cases the US dollar. The fixed rate, in practice a currency band, is maintained by tying the domestic money supply to the in- and outflow of foreign currencies. After stabilization of the domestic currency, i.e. during an initial boom phase, the domestic economy experiences a first inflow of foreign capital, resulting in an expansion of the domestic money supply and rising prices, both in domestic currency and dollars. However, if the foreign capital inflow is not forthcoming, the domestic economy experiences a domestic currency cost crunch, which expresses itself as rising prices valued in the domestic currency but not in dollars. One response to this situation is a devaluation of the domestic currency; another is that the government offers higher returns for investments in order to attract foreign capital inflow. The latter is very costly since the government has to finance the interest paid on investments; depending on the state's financial capacities to finance investment returns, the final result will probably be devaluation, but at the cost of having far fewer resources than before the attempt to defend the domestic currency (Woodruff 2005).

After the introduction of the rouble currency band, inflation declined and the real exchange rate increased, causing the rouble to appreciate against the dollar and making it more attractive for foreign capital investors. However, capital only entered the country after mid-1996, when Boris Yeltsin strong defender of the market course, was re-elected Russian president. Before that date the political uncertainties were too high. Throughout 1996 and 1997 foreign liabilities grew strongly, in fact dramatically. There was a general development towards dollar-denominated liabilities; for instance many local, regional, and federal governments sought to borrow in dollars rather than in roubles, for which the interests where higher. Also, many exporters borrowed abroad, predominantly in the energy sector. (Woodruff 2005). By that time, inflation had become only a minor problem (Woodruff 2000). The currency band provides evidence, however, that the Russian state did not want to give up the facility to attract investments from abroad in order to drive production and economic growth. Although the policy is risky and only works well when investments continue to flow, in the opposite case the lack of investments jeopardizes the stability of the domestic currency.

The growth of the importance of the dollar in the Russian economy increased the importance of the domestic market for export-oriented enterprises, such as oil and gas companies. The rising purchase power of Russians made it easier for those companies to transform their infrastructure for the supply of foreign markets, whereas previously the infrastructure was geared towards the supply of the domestic market. Clearly, however, not all enterprises benefited from the general appreciation of the rouble during the existence of the currency band. Producers that competed with imports were not very well equipped to adjust to the cheaper products from abroad, especially due to the nominal price-stickiness of many domestic products; producers' costs could not be reduced due to prices being nominally rigid. The state sought to compensate for this problem by intervening in the market, one measure being the introduction of protective tariffs. But, in the light of the general cost pressures in the domestic Russian economy, many producers found their way to monetary surrogates (Woodruff 2005). In the forestry industry, state interventions also found their expression in the involvement of public authorities in local and regional forestry enterprises in the form of exchanging company shares against tax liabilities. Furthermore, state monopolies contributed to infusing money into the quasi-private forest sector (Carlsson et al. 2000). From the perspective of the private enterprises involved, this led to dependencies and maladaptations with respect to introducing market-oriented production processes. Thus, these enterprises lacked important prerequisites for their own sustainability.

A general cost crisis can also be described as a recessionary tendency in the Russian economy. This recession resulted in a conflict of interests between enterprises whose costs and sales were denominated in roubles and export-oriented enterprises for whom the dollar rate was far more important. Furthermore, strains were put on the relations between debtors and creditors. The response to this crisis was a focus on a variety of possible alternatives to a devaluation of the rouble. Alternatives included market interventions to create a downward flexibility of costs in the domestic economy and forbearance in the enforcement of debt repayment and contract fulfilment. The reason for not giving preference to an immediate devaluation has to be seen in the light of the constellation of interests surrounding exchange rates and prices Moreover, the Russian federal government did not have a majority that favoured a devaluation of the rouble; there was a strong desire to support the Russian banking system during the crisis existed. Russian commercial banks had accumulated large rouble deposits, financed with dollar liabilities. To save the banks' solvency the devaluation was delayed (Woodruff 2005). Preserving the fixed exchange rate was also seen as a matter of credibility and fostering market confidence in the eyes of investors (Pinto et al. 2004).

It becomes apparent what the occurrence of the high-risk case of an investment stop meant for the participants of the Russian economy: a massive loss of trust by those who were dependent on the rouble and ignorance on the part of those who were more independent of it. This conflict of interests led to a rapid fragmentation of the Russian monetary system. Monetary surrogates began to dominate inter-enterprise transfers and indeed serviced an alternative system of payment based on trade credit. In addition, Russian governments, local and national, issued surrogates, which created huge problems at a later stage, when the widespread demonetization of the Russian economy led to substantial fiscal deficits due to a reduction of government taxes (Woodruff 2005). The forest sector as a whole was heavily affected by the increase in non-monetary transactions. Many, even large, enterprises in the sector sold only a small share of their production for real money, an estimated ten per cent. Among non-monetary transactions one can also count activities involving so-called relational capital investments; these include performing services for local authorities or incorporating these authorities into the ownership structure of forestry enterprises in order to, for example, offset tax liabilities or negotiate privileges (Carlsson et al. 2000). Issuing monetary surrogates required further borrowing on the government's part, especially in order to meet debt repayment obligations for which surrogate means would not work. Eventually, devaluation became a reality in August 1998 (Woodruff 2005). The crisis had indeed been very costly. The Gross Domestic Product (GDP) of Russia had fallen by four point nine percent in 1998, the annual inflation rate in

December of that year reached 84 per cent, compared to a target of eight per cent, and some $30 billion in foreign exchange had been used to protect the rouble from devaluation between October 1997 and August 1998, when the decision to float the rouble was made. Since then the Russian economy has made striking progress: for example, only a year later, in 1999, the GDP had already grown by five point three percent and the inflation rate had fallen to less than 40 per cent (Pinto et al. 2004).

Monetary Instruments

In the early 1990s the instruments which the Central Bank of Russia had at its disposal in its monetary policy were limited to the control of directed credits (to specific enterprises or industrial sectors) as well as reserve requirements to control monetary aggregates. All commercial banks that had been created since the split of Gosbank's function were subject to strict reserve requirements. A certain proportion of the banks' reserves had to be kept in accounts with the central bank. As inflation soared in 1992, the reserve requirements were increased to a larger percentage of the banks' reserves. The enforcement of the reserve requirement was rather poor, however. Furthermore, the rouble assets in the banks were quite unevenly distributed. A few banks had a large share of all the assets, while a large group of banks had a comparatively small share. Any case of further tightening of reserve requirements would mean difficulties for some banks. The most significant measure which hampered the Central Bank's monetary policy, however, was its allowing commercial banks to obtain short-term credit by overdrawing their reserve accounts. Regional managers of the central bank had virtually automatically provided these overdrafts to the banks in their jurisdiction; even though the charge was twice the refinancing rate, the charge in real terms was still negative in the light of the high inflation in 1992 (Baliño et al. 1997). Thus, initially, the general idea of orthodox economic policy prevailed – fuelling the economy with capital to maintain or to raise production – rather than keeping calm and focusing on currency stability.

After the banking system developed further and economic conditions stabilized around 1994 and 1995, the Central Bank of Russia introduced new monetary instruments. These included the active use of interest rates to reduce the demand for credit from the Central Bank. Market-based instruments were considered after commercial banks had been stabilized to the extent that they could take part in financial market activities. The Central Bank introduced a market-based credit facility that was capable of providing short-term liquidity to the commercial banks (Baliño et al. 1997). In addition, many banks made use of the possibility to trade credit on an inter-bank market, which was an important mechanism as it distributed liquidity from those banks with a surplus to those banks in need of cash (Furfine 2001).

The Central Bank of Russia continuously modified the effectiveness of its reserve requirements. Rates were changed from time to time to allow greater flexibility and to keep costs down for all commercial banks. In addition, compliance with the reserve facility was tightened. Overall, the Central Bank has been fairly successful in developing monetary instruments which inject liquidity into the Russian domestic economy. However, the credit auctions and reserve requirements which provided liquidity to banks were not suitable for absorbing liquidity. Measures were undertaken to manage a surplus of money: the Russian Ministry of Finance sold treasury bills in excess of its financial needs and the Central Bank started to auction deposits. However, the latter measure was not as successful as hoped because the placing of the deposits occurred at an interest rate determined by the Central Bank and not by the market; the rate offered by the bank was significantly below the yield of other options (Baliño et al. 1997).

The implementation of monetary policy by the Central Bank of Russia mainly took the form of using monetary aggregates as a tool, focusing on money supply via the reserve requirements and deposit auctions. Although the target of that policy before 1995 was to decrease the inflation rate, after 1995 the objective was exchange rate stabilization (Esanov et al. 2004). This is due to Russia establishing an exchange rate programme starting in 1995 (Woodruff 2005). Nevertheless, the use of monetary aggregates as an instrument to conduct monetary policy appears in contrast to the experience of other emerging markets, where interest rate policies rule. Given the increasing credibility of the Central Bank of Russia, the development of domestic financial markets, and the policy reforms undertaken in late 2002, interest rate policy coupled with inflation targeting in a floating exchange rate regime should in time be successfully implemented (Esanov et al. 2004).

A fact supporting this argument is that Russia experienced 3500-fold inflation at the same time as a 1500-fold increase in monetary supply. In orthodox[1] monetary policy the control of the growth of monetary aggregates helps to control inflation. The Central Bank of Russia follows this policy. The money supply here is viewed as an exogenous parameter; i.e. it is controlled from outside the economy. Inflation, then, is seen as the result of an excessive money supply. On the other hand, heterodox[2] monetary theory presupposes that the interest rate, which is based on the price of credit, is an effective means to control inflation (Vymyatnina 2006).

---

1   Bofinger (1996) describes orthodox economic policy approaches as being consistent with a strong focus on controlling the money stock by using fiscal and monetary policy. This can be considered inadequate if the demand for money – especially the case in a transition economy – is unstable. Though, modern central banking is based on orthodox theory, founded on the assumption that a central bank – independent from the state – can exogenously control money in the economy (van Lear 2003).

2   Heterodox economic policy advocates that the central bank's ability to control money is limited because of the endogenous supply of money in the economy. Thus, heterodox policy aims to emphasize the role of the state and considers income and employment policies, arguing that inflation is the result of income distribution conflicts, leading to the use of non-monetary exchange media. (van Lear 2003) Thus, Bofinger (1996: 669) calls heterodox approaches a commitment technology, stabilizing incomes to allow a gradual disinflation.

In mature economies the source of an endogenous money supply is the demand for credit by enterprises to fund their industrial capital and investments. This is different in economies in transition. Here, one can distinguish between an old and new industrial enterprise sector: a new corporate sector operating within the logic of the market and an old sector still operating under the logic of a planned economy. The latter type of enterprises might in many cases operate completely unprofitably. The reasons why these enterprises are still running include their often being the only source of employment in a locality. Local authorities, therefore, are eager to pressure for the continuation of credit provision to those old-type enterprises, with obvious effects on the monetary system (Vymyatnina 2006).

Thus, in today's Russia there are several sources of endogenous monetary growth. One stems from credit demanded by market-operating enterprises, a second from credit passed on to socially important enterprises. Finally, a third source is barter. Barter introduces money that is completely endogenous economically, beyond any control (Vymyatnina 2006). The use of barter was greatest during the height of the currency crises in August 1998, but still stands at about 10 per cent (Sutela 2003). Evidence suggests that the money supply in Russia is endogenous. As said before, Russian monetary policy is based on orthodox theory, which assumes that the money supply is exogenous. Hence, the Central Bank cannot effectively and predictably control the growth of monetary aggregates. Besides, controlling monetary aggregates, which involves reserve requirements for banks, is excessive and might well prevent more effective economic development. Where monetary policy is concerned, the operational goal of the Central Bank is to set the growth rate of the monetary base, since this is the parameter conceived to be completely under its control. Direct control measures that target commercial banks' lending are used. These measures include the control of liquidity and solvency of banks. Regulations state that banks cannot give too much credit to enterprises and also include prescribed interest rates (Vymyatnina 2006).

There is still a tension in monetary policy between lowering inflation and slowing down the appreciation of the exchange rate as part of favourable trade conditions. So far, the Central Bank of Russia has preferred to slow down the appreciation of the exchange rate to provide better conditions for the domestic export industry. This forces the Central Bank to buy foreign currencies from the market, thereby causing inflation. One indicator for this policy is the persistent inflation rate – still around 10 to 12 per cent – which is moderately higher than planned. The apparent policy suggests that interest rate control is not used as an instrument of direct inflation control. As such this is not surprising in Russia, where the present regime of monetary control mechanisms took a long time to be established and to become stable. However, due to the apparent tension and trade-off between exchange rate stability and inflation control, this move towards an interest rate control mechanism should be made (Granville and Mallick 2006). This would necessitate a change in the attitude of Russian monetary policy. As mentioned, the Central Bank of Russia follows an orthodox path. Using the interest rate as an effective control mechanism is the main postulate of heterodox monetary theory (Vymyatnina 2006).

In the light of the requirements for a suitable macroeconomic policy as presented above, the move towards a policy based on heterodox monetary theory would more coincide with the need to maintain trust in an economic area's currency. So far, the priority has been to create favourable conditions for economic growth, at the expense of controlling the currency value. Favouring economic growth via the exchange rate and accepting inflation, albeit moderate, appears to be the prevailing approach in Russia – one which seems to focus on short-term growth rather than long-term trust maintenance.

Conclusion

Certain policy developments suggest that the investment climate in Russia is improving. This is firstly due to the much better fiscal and monetary policies. Early fiscal policy erased the trust that existed in the currency through irresponsible budget allocations, leading to high inflation which could not be controlled by the instruments defined in monetary policy. Furthermore, monetary policy was for a long time too dependent on political authority, leading to a practice where budget holes where financed through central bank credit, which in turn fuelled consumer price inflation.

Many measures have been undertaken to increase the trust in the rouble. These include a tightened fiscal policy, which, as shown, has contributed to producing sizable budget surpluses since about 2000. Around the same time, the federal tax code was reformed and simplified, making it more efficient and easier to capture revenues and windfall profits from oil and gas exports. There are still problems with Russian monetary policy, where there is a perceived contradiction between policy goals. That means either keeping the rouble exchange rate below market prices in order to support exports of domestic industries or decreasing inflation. In practice this has led the Central Bank aiming for a deceleration of rising inflation, which continues at about 10 to 12 per cent per annum.

In addition, financial policy has somewhat changed since 1 July 2006; the rouble is now freely convertible. This has been a political step to indicate that the Russian economy and currency are now strong enough. Naturally, it is hoped that this move will help to increase the level of foreign investment. Thus, it could bring important changes to the corporate sector, while the private sector might experience lower prices for imported goods due to the previously undervalued rouble. Russians themselves are often still suspicious about the stability of the currency and the security of bank deposits. Many do not keep their savings in bank accounts. The ghost of the crash in 1998 is still haunting the country.

Although developing well at the moment, the Russian economy does not have any problems in absorbing the increasing volume of money (M2) in circulation that is denominated in roubles. Even though the current inflation rate is still comparatively high, it does not seem to provide much of a constraint on economic growth at the moment. It does, therefore, provide an argument for those who favour the use of the exchange rate to drive economic development. But this should be seen as short-termism. Quick and easy growth of the economy does come at a cost – after all, 10 to 12 per cent inflation per year is also a 10 to 12 per cent loss of trust per year. A long-term sustainable development perspective should be centred on low inflation, high trust, and be designed independently of growth objectives.

Thus, there are risks associated with the present development strategy. There is the long-term risk of not giving sufficient credence to the importance of trust in the currency and the short-term risk of having to deal with a decreasing level of investments from abroad. Although the latter does not seem apparent, recent arbitrary actions by governmental authorities, ranging from the Yukos affair to Shell's Sachalin case and the instrumental use of the energy infrastructure for political purposes, do not send out positive signals to investors. That such cases which threaten the security of investments are not confined to the energy sector but affect other resource sectors as well, such as forestry, is shown by the other chapters in this book, although the state has admittedly a differing role.

I have attempted to show that governing money badly threatens sustainable development in the economy and society. Firstly, as a consequence of bad governance investments are lacking. As noted, macroeconomic policy has for a decade failed to provide sufficient trust to motivate investments. Secondly, the use of barter has for a long time been quite high, although it is estimated to be a tenth of what it was at the height of the economic crisis in 1998 (Sutela 2005: 16). This should be considered an average figure; it is may well be higher in some regions, industries, and particular enterprises elsewhere in Russia. Accordingly, economic policy has had its difficulties being effective. Governing the economy and sustaining development requires cooperation, which in turn is based on trust. Using money as a general medium of exchange is a form of cooperation. Using the same currency within an economic area allows governance, as governing the value and stability of money in such a way that monetary alternatives, such as barter, are not interesting, establishes the required trust. Otherwise, macroeconomic policy fails to provide a structural foundation for future governance. In addition to having a cooperative social dimension, trust has a temporal dimension that affects confidence and expectations into the future. On the basis of the existing inflation and the careful dealing with banking facilities one can conclude that full confidence has not been restored as yet. It is these social and temporal dimensions of trust in the economy, which, I believe, are most important here; they are more important than any "economic" dimension of trust for the purpose of achieving, say, economic competitiveness.

The ideas presented in this chapter also illustrate well the nature of politics as the governing system in society. Just like money is circulating in the economy and causes problems for sustaining the economy when absent, the political system in Niklas Luhmann's theory of social systems is of a cyclic nature. In a self-reproducing political system, governance could be seen as the element which has to be reproduced in order to ensure the continuity of the political system. Thus, future ability to govern requires good governance at present. If not, the power to steer might be lost and sustainable development jeopardized.

References

Ahrend, Rudiger (2004), 'Accounting for Russia's Post-Crisis Growth', OECD *Economics Department Working Papers*, No. 404 (Paris: OECD Publishing).

Ahrend, Rudiger (2006), 'How to sustain growth in a resource based economy? The Main Concepts and Their Application to the Russian Case', *OECD Economics Department Working Papers*, No. 478 (Paris: OECD Publishing: Paris).

Aizenman, J. and Pinto, B. (eds), *Managing Volatility and Crises: A Practitioner's Guide* (Washington D.C.: The World Bank).

Aleksashenko, S. A. (2000), 'The Monetary Policy: Is Normalization Really Achieved?', *International Monetary Fund-Conference "Investment Climate and Russia's Economic Strategy", Moscow, April 5 – 7, 2000*, URL https://www.imf.org/external/ pubs/ft/ seminar/2000/invest/pdf/alek2.pdf, 25.05.2007.

Baliño, Tomás J. T. (1998), 'Monetary Policy in Russia', *Finance & Development*, 35, 4, pp. 36–39.

Baliño, Tomás J. T., Hoelscher, David S. and Horder, Jakob (1997), 'Evolution of Monetary Policy Instruments in Russia', *IMF Working Paper*, WP/97/180, (Washington D.C.: International Monetary Fund).

Bofinger, Peter (1996), 'The Economics of Orthodox Money-Based Stabilisations (Ombs): The Recent Experience of Kazakhstan, Russia and the Ukraine', *European Economic Review*, 40, pp. 663–671.

BOFIT (2004), Russia. Harell, Timo (ed.) *BOFIT Weekly*, 4/23.1.2004, (Helsinki: Bank of Finland Institute for Economies in Transition (BOFIT).

BOFIT (2007), *BOFIT Venäjä-tilastot*, (Helsinki: Bank of Finland Institute for Economies in Transition (BOFIT), URL http://www.bof.fi/bofit/seuranta/venajatilastot/, 21.05.2007.

Carlsson, Lars, Olsson, Mats-Olov and Lundgren, Nils-Gustav (2000), 'If Money Only Grew On Trees: The Russian Forest Sector in Transition', *The Forestry Chronicle*, 76, 4, pp. 605–610.

Clesse, A. and Zhurkin, V. (eds). *The Future Role of Russia in Europe and in the World* (Luxembourg: Luxembourg Institute for European and International Studies).

Daly, Herman (1992), 'Allocation, Distribution, and Scale: Towards an Economics That is Efficient, Just and Sustainable', *Ecological Economics*, 6, pp. 185–193.

Dudarev, Grigory, Boltramovich, Sergey and Efremov, Dmitry (2002), *From Russian Forests to World Markets – A Competitive Analysis of the Northwest Russian Forest Cluster* (Helsinki: The Research Institute of the Finnish Economy (ETLA)).

Dudarev, Grigory, Boltramovich, Sergey, Filippov, Pavel and Hernesniemi, Hannu (2004), '*Advantage Northwest Russia: The New Growth Centre of Europe?* (Helsinki: The Finnish National Fund for Research and Development (Sitra)).

Esanov, Akram, Merkl, Christian and de Souza, Lúcio Vinhas (2004), 'A Preliminary Evaluation of Monetary Policy Rules in Russia', *Kiel Working Paper*, No. 1201 (Kiel: Kiel Institute for World Economics).

Furfine, Craig (2001), 'The Interbank Market During a Crisis', *BIS Working Papers*, No. 99 (Basel: Bank for International Settlements).

Granville, Brigitte and Mallick, Sushanta (2006), 'Does Inflation or Currency Depreciation Drive Monetary Policy in Russia?', *Research in International Business and Finance*, 20, 2, pp. 163–179.

Holopainen, Päivi, Ollonqvist, Pekka and Viitanen, Jari (2006), 'Factors affecting investments in Northwest Russian forest sector and industry', *Working Papers of the Finnish Forest Research Institute*, No. 32 (Helsinki: Finnish Forest Research Institute (METLA)).

Jalava, Janne (2003), 'From Norms to Trust: The Luhmannian Connections Between Trust and System', *European Journal of Social Theory*, 6, 2, pp. 173–190.

Jordan, G. J. and Fortin, M.-J. (2002), 'Scale and Topology in the Ecological Economics Sustainability Paradigm', *Ecological Economics*, 41, pp. 361–366.

Komulainen, Tuomas (1999) 'Currency Crisis Theories – Some Explanations for the Russian Case', *BOFIT Discussion Papers*, No. 1/1999 (Helsinki: Bank of Finland Institute for Economies in Transition).

Kooiman, Jan (1993), *Modern Governance: New Government-Society Interactions* (London: Sage).

Lewis, J. David and Weigert, Andrew (1985), 'Trust as a Social Reality', *Social Forces*, 63, 4, pp. 967–985.

Luhmann, Niklas (1970), *Soziologische Aufklärung: Aufsätze zur Theorie sozialer Systeme* (Opladen: Westdeutscher Verlag).

Luhmann, Niklas (1984), *Soziale Systeme: Grundriß einer allgemeinen Theorie* (Frankfurt am Main: Suhrkamp Verlag).

Luhmann, Niklas (1986), *Ökologische Kommunikation* (Opladen: Westdeutscher Verlag).

Luhmann, Niklas (1988), *Die Wirtschaft der Gesellschaft* (Frankfurt am Main: Suhrkamp Verlag).

Luhmann, Niklas (1997), *Die Gesellschaft der Gesellschaft* (Frankfurt am Main: Suhrkamp Verlag).

Luhmann, Niklas (2000), *Vertrauen*, 4th Edition (Stuttgart: Lucius & Lucius).

Mashkina, Olga (2006), 'The Russian Forest Industry in Transition: Historical-Institutional Perspective', *XIV International Economic History Congress, Helsinki, 21–25 August 2006*, URL http://www.helsinki.fi/iehc2006/papers3/Mashkina.pdf, 25.05.2007.

Orlowski, Lucjan T. (1997), 'Russia's Economic Stability: Recent Evidence and Policy Implications', in Clesse, A. and Zhurkin, V. (eds).

Pinto, Brian, Gurvich, Evsey and Ulatov, Sergei (2004), 'Lessons from the Russian Crisis of 1998 and Recovery', in Aizenman, J. – Pinto, B. (eds).

Rhodes, R. A. W. (1996), 'The New Governance: Governing without Government', *Political Studies*, XLIV, pp. 652–667.

Sutela, Pekka (2003), *The Russian Market Economy* (Helsinki: Kikimora Publications).

Sutela, Pekka (2005), 'Will Growth in Russia Continue?', *Bank of Finland Bulletin*, 4, 79, pp. 12–20.

van Lear, William (2003), 'Implications Arising from the Theory on the Treasury's Bank Reserve Effects', *Journal of Post Keynesian Economics*, 25, 2, pp. 251–261.

Vymyatnina, Yulia (2006), 'How Much Control does Bank of Russia have Over Money Supply?', *Research in International Business and Finance*, 20, 2, pp. 131–144.

Walter, Stefan (2008), 'Understanding the Time Dimension in Resource Management', *Kybernetes*, 37, 7, pp. 956–977.

Woodruff, David M. (2000), 'Rules for the Followers: Institutional Theory and the New Politics of Economic Backwardness in Russia', *Politics & Society*, 28, 4, pp. 437–482.

Woodruff, David M. (2005), 'Boom, Gloom, Doom: Balance Sheets, Monetary Fragmentation, and the Politics of Financial Crisis in Argentina and Russia', *Politics & Society*, 33, 1, pp. 3–45.

Chapter 4

Russian Forest Regulation and the Integration of Sustainable Forest Management

Minna Pappila

Regulating Sustainability

If a little green alien were to carefully read our international environmental agreements and declarations and take them literally and seriously, it would assume that humankind is already moving rapidly towards sustainable life on Earth. The reality is, however, not as encouraging. Some minor battles have been won, but much remains to be done. For example, biodiversity, fresh water resources and the fertile soils of the world are declining and climate change is becoming more evident and threatening.

The Russian Federation is the site of one-fifth of the world's forests and has signed all important international documents concerning forests and sustainable development, e.g. the Convention on Biological Diversity (CBD), the non-legally binding Forest Principles,[1] and the St. Petersburg Declaration.[2] Russia has also taken part in developing criteria and indicators for sustainable forest management within the Montreal Process and Ministerial Conferences on the Protection of Forests in Europe. The Russian Federation has thus, like many other states, committed itself to the sustainable development of forests.

It is indisputable that Russia hosts huge forest areas and the last vast intact forest landscapes of Europe (Yaroshenko et al. 2001); it is equally true that Russia has a history of forest decline in some regions (Weiner 2005, 209), insufficiency of forest regeneration and severe regional overharvesting (Pisarenko and Strakhov

---

1 Forest Principles i.e. Non-legally Binding Authoritative Statement of Principles for a Global Consensus of the Management, Conservation and Sustainable Development of All Types of Forests (approved in Rio de Janeiro in 1992). States have not succeeded in concluding a binding agreement on forest protection. The United Nations Forum on Forests (UNFF) carries on the global dialogue on forest issues and possible global convention on forests.

2 St. Petersburg Declaration (2005) is part of the ENA FLEG (Forest Law Enforcement and Governance) process to take action to address illegal logging and associated forest crimes.

1996, 29). The majority of the remaining intact forest landscapes have been protected mostly because of their remoteness, not as part of a deliberate protection policy. According to the Russian Red Data Book of endangered species, 40 per cent of endangered plants and 45 per cent of endangered animals are closely linked to forest ecosystems and their future is thus highly dependent on the level of biodiversity protection in Russian forests (Russia's Report 2003). Hence, there is an evident call for an ongoing process of sustainable forest management in Russia, as elsewhere in the world, too.

Since the collapse of the Soviet Union, three Forest Acts have already been enacted in the Russian Federation – one in 1993, another in 1997, and the latest at the end of 2006. A comprehensive reform of the old 1997 Forest Code started in 2002 and the new Forest Code was enacted at the end of 2006. There have been considerable organizational changes in the forest sector in the years since the collapse of the Soviet Union, but practical, down-to-earth regulation on felling has not been amended as much as the structure around it, e.g. private enterprises, leasing, permits, surveillance, authorities. The stability of the felling rules has probably guaranteed that forestry has somehow survived the transition period.

The Forest Code is the most important law on the use of forests in Russia. The scope of application of the Code covers all the forests in the Federation. It even includes some rules concerning forestry in nature protection areas, e.g. national parks, although yet the main regulations on nature protection areas are in a special law on specially protected areas. Logging is allowed in some parts of protection areas, however, and the permitted forms of logging have been specified in the Forest Code. More detailed provisions on logging and leasing are set out in numerous special enactments.

The aim of this chapter is to evaluate to what extent Russian governmental forest regulation reflects the international regulation and its norms of sustainable forest management. Here, the drafting process of the new Forest Code, and the outcome, i.e. the new law on forest use, are assumed to reveal the course of the contemporary forest regulation. Before analysing the process and the law, the concepts of sustainable development, sustainable forest management and the integration principle will be defined. Then both the law-making process and the outcome will be assessed. Forest policy as a possible panacea for problems related to sustainable forestry will be discussed in the conclusion of this chapter.

This contribution concentrates on state regulation, even though Russian and global forest governance includes several other forms of regulation, such as state regulation, international regulation (e.g. the CBD), inter-governmental regulation (e.g. the Montreal Process) and self-regulation (e.g. forest certification schemes). It is the national forest regulation, however, that sets the framework for other kinds of regulation and governance in a given country.

Central Concepts

Sustainable Development

This chapter is based on the concepts of sustainable development, sustainable forest management and the integration principle. Sustainable forest management is part of sustainable development and one of the numerous means to pursue the aim comprehensively.

There is no consensus on the content of sustainable development. Nevertheless, while sustainable development is an elusive goal, it is also a widely accepted legal principle of international and, in many instances, national environmental law. According to the most often quoted definition, sustainable development means development that meets the needs of the present without compromising the ability of future generations to meet their own needs (WCED 1987, 8). Sustainable development should – according to the integration principle – nowadays be part of almost all human activities and policies, for example, agricultural, energy, forest, land use, science, transportation and waste policies, as well as consumption and production patterns and institutional arrangements.

Sustainable development is typically divided into ecological, economic, social and cultural sustainability, which sometimes go hand in hand yet in some cases collide. In some instances, it is easy to see both ecological and economic benefits in the process of sustainable development.[3]

While ecological sustainability cannot always be distinguished separated from other aspects of sustainability, it is the basis for other forms of sustainability and also the starting point for this contribution. The understanding of sustainable development in this chapter could be called 'strong sustainable development', which means, for example, that environmental protection is a precondition of economic development (Baker et al. 1997, 13–15). Without ecological sustainability also economic sustainability will fail in the long run. Ecological sustainability can be divided into numerous sub-categories, and protection of biodiversity is one of them. Article 2 of the CBD defines biodiversity as 'the variability among living organism from all sources including, inter alia, terrestrial, marine and other aquatic ecosystems and the ecological complexes of which they are part; this includes diversity within species, between species and of ecosystems'. The loss

---

3   This can be the case with modernizing paper mills; new equipment and production methods that save raw material and energy can lead to economic benefits. There will also be less air and water pollution and less wood is needed to get the previous amount of paper. In cases where we protect the habitat of an endangered species in a managed forest, someone will have to bear the economic burden; this is usually the owner, the leaseholder or the state. In this case it can be more difficult to get a win-win case, unless the forest owner gets better compensation for logs originating from his or her ecologically well-managed forests; in that case the consumer is usually expected to be willing to pay more for an ecologically friendlier product.

of biodiversity is accelerating and is for the most part irreversible (Millennium Ecosystem Assessment 2005).

Instead of trying to define sustainable development precisely, it is more fruitful to see sustainable development as a social and political construct that is changing constantly but that has certain underlying goals (Baker et al. 1997, 6). It can be claimed that the construct includes everything and nothing and it is relatively easy to use it as mere political rhetoric or a slogan. Yet, it could also be used as an intensive for continuous improvement of human activities. No country or sector of the economy (e.g. forestry, farming, tourism, housing, fishing, mining, energy production, or apparel) can claim to be totally sustainable if we take all aspects of sustainability into account.

A general principle such as sustainable development has to be divided into sub-goals, e.g. sustainable forest management, in order to be implemented and to assess the success of implementation. Evaluating current implementation processes makes it possible to assess the trend of development. In the present case, this means that by looking at the forest law reform we can assess whether the Russian forest regulation is on an ecologically sustainable path or not. However, at the same time we must remember that sustainable forest management alone does not lead to ecologically sustainable forests; other forms of land use and pollution have an effect on forests. In order to protect overall biodiversity there also has to be nature conservation areas where forestry is forbidden altogether. No matter how ecologically sustainably forests are managed, there are always species that require large enough tracts of natural forests to survive.

Sustainable Forest Management (SFM)

Sustainable forest management, as it is currently understood, sprang up during the last decades of the twentieth century, mostly in the 1990s, in parallel with sustainable development discourse (Adamowitz and Burton 2003, 44).[4] For decades, sustainability in forestry meant sustained timber yield, i.e. economic sustainability: wood consumption should not exceed wood production in the long term. Neither biodiversity loss nor social and cultural consequences were part of this

---

4   The change that took place in forest management during the past three decades can be seen as a part of a general cultural change, at least in the western parts of the world. Earlier natural resources management was based on the idea of objective expert knowledge and the main aim of forest management was fibre production for industry. Social factors were considered problems, not a potential source of goals and values to be taken into account. At the same time, nature conservation was founded on utilitarian values, i.e. the greatest good for the greatest number, emphasizing science and the use of deterministic scientific models that would predict changes in nature (Schelhas 2002, 18). Russia has been accused of still relying heavily on technology and having a very utilitarian attitude towards nature (Tynkkynen and Massa 2001, 15; Weiner 2005, 227), and thus perhaps this modernization of forest management and biodiversity protection has not yet reached Russia.

kind of sustainability. Ecological issues were considered only in relation to nature protection areas. Gradually, sustainability has become more plural, and ecological sustainability is considered – at least in principle – an equal part of SFM.

Currently, sustainable forest management could be understood as forest management that preserves the health of forest ecosystems while taking into account the ecological, economic, social and cultural values of the forests for present and future generations (Duinker and Trevisan 2003, 857). Work et al. (2003, 953) describe SFM as 'a broad, process best expressed as ongoing framework for development rather than a state to be achieved and forever proclaimed'. The Ministerial Conference on the Protection of Forests in Europe defined SFM as 'the stewardship and use of forests and forest land in a way, and at a rate, that maintains their biodiversity, productivity, regeneration capacity, vitality and their potential to fulfil, now and in the future, relevant ecological, economic and social functions, at local, national and global levels, and that does not cause damage to other ecosystems' (MCPFE 1993). The area set aside for protected forests (e.g. nature protection areas) and water protection zones, the number of endangered forest species and the area of forest land exposed to specific air pollutants are some of the indicators of the sustainability level of forests defined in the criteria and indicators of the Montreal Process. In light of the above-mentioned definitions of SFM and its interest in ecologically sustainable development, the chapter will proceed to focus on ecological values in general and biodiversity protection in particular.

Parts of the content of SFM is found, for example, in international agreements and soft law, but it is not possible to cover the concept through an exhaustive list of elements. Rather, one has to choose certain criteria and use them as a yardstick for evaluating sustainability.

Clearly, one aim of SFM is to combine forest management and forest biodiversity protection instead of separating forestry and nature protection. Safeguarding biodiversity can be divided into three parts: conservation (protection areas), sustainable use (e.g. responsible forestry) and non-degradation (e.g. protection against air and water pollution) (Kokko 2003, 280). Modern nature protection requires a much more extensive toolbox than in the first decades of nature protection in the early twentieth century, when the establishment of nature protection areas was the main goal and means of nature protection. Nowadays, the ideal is to integrate biodiversity protection into all spheres of human activity. In addition to integration into many fields, an array of biodiversity protection instruments have been introduced in several countries ranging from national parks and habitat protection to various forms of voluntary protection, e.g. subsidies, agreements and auctions for the best protection areas and activities. SFM can be seen as one means of biodiversity protection though sustainable use of forests.

It is possible to recognize a gradual move towards more plural sustainability in the international forest management discourse, but it is difficult, if not impossible, to determine the exact content of the concept of sustainable forest management. This varies from country to country and from one stakeholder to another. Stakeholders interpret and implement it differently: states, ENGOs (environmental non-governmental organizations) and industrial groups have created their own guidelines for sustainable forest management (Adamowitz and Burton 2003, 46). It is clearly a concept that is being constructed again and again in being used and discussed. Continual changes in our knowledge and values cause the concept to change accordingly. Gradually these changes should also reach the practical forest management level and lead to continuous improvement.

Sustainable development and SFM do not necessarily have the same meaning in Russia as they have in the Western discourse. Russia has its own long tradition of forestry and a "A Great Power has no problems" attitude towards environmental issues (Tynkkynen 2005, 290). It has been said that sustainable development was initiated in Russia from above in the form of a presidential decree and without any public discussion (Yanitsky 2001, 42). This is clearly an erroneous starting point and in such a case sustainable development easily remains so much political rhetoric. On the other hand, at least in the sphere of forest protection and SFM, many ideas have been introduced in Russia by civil society, for example, the importance of the protection of old-growth forests and responsible forestry through forest certification (see e.g. Chapter 9 in this volume). Russian ENGOs have raised awareness, with help from transnational civil society. They have managed to convince at least some of the stakeholders that new means of forest protection are necessary. Even if this is partly forced sustainability, it might affect values in the course of time. In addition, even if one were to consider the concept of sustainable development irrelevant in Russia, there have been elements of sustainable development in Russian forestry and forest protection for some decades, e.g. water protection zones, specially protected forest tracts in managed forests, and different kinds of nature protection areas. The idea in this contribution is to ascertain whether certain elements of SFM found in the Forest Principles exist in the Russian forest regulation, not to find out how SFM has been perceived in the Russian forestry discourse, although the discourse on forest legislation has been used as research material here.

The Integration Principle

> [I]f environmental factors are not taken into consideration in the formulation and implementation of the policies that regulate economic activities and other forms of social organization, a new model of development that can be environmentally and socially sustained in the long term cannot be achieved. (Liberatore 1997, 107)

An important way to further operationalize sustainable development and sustainable forest management is to change legislation accordingly. Integration of environmental policies into all sectors of society is a precondition for sustainable development. Sustainable development is not only a matter of environmental legislation; rather, it should be integrated into all fields of legislation. Integration is one of the leading principles in environmental policy of the EU today, and the integration principle is also part of many international agreements and declarations. Article 10(a) of the CBD urges states to integrate consideration of the conservation and sustainable use of biological resources into national decision-making. This means, for example, that states should improve the co-ordination of decision-making between administrative sectors so that they do not have contradictory objectives. States should also co-ordinate legislation so that different acts will not conflict and thus hinder the implementation of environmental laws and policies (Glowka et al. 1994, 58–9).

Integration may be considered to include several dimensions, which may be issue-specific, sectoral, spatial and temporal, organizational, distributive and ethical (Liberatore 1997, 113). On the one hand, environmental problems should be taken seriously in different sectors, such as forestry, agriculture, transport and trade. On the other, according to the issue approach, problems such as water pollution should be considered in several sectors simultaneously to develop a comprehensive policy. Here only the issue dimension will be discussed in the framework of forest regulation. This means considering how selected issues of ecologically sustainable development have been integrated into forest legislation.

The integration principle is also part of the Forest Principles, which date back to the Rio Summit of 1992: the participating states could not agree on concluding a binding agreement on forests and instead adopted non-binding and rather general principles. Nonetheless, these are the only globally agreed criteria for sustainable forest management. Since Rio, the global intergovernmental forest dialogue has continued in the UN-based IPF, IFF and Forest Forum.[5] Articles 3(c), 6(b) and 13 (d) of the Forest Principles aim at intersectoral integration by stating that all aspects of environmental protection and social and economic development as they relate to forests and forest lands should be *integrated* and comprehensive and that forest conservation and sustainable development policies should be *integrated* with economic, trade and other relevant policies (emphasis added).

In addition to the integration principle, the Forest Principles embrace also other widely accepted principles and issues, among these sovereignty over natural resources, the participation principle and the rights of indigenous people among others. These are also relevant from the point of view of sustainability but are not examined here, because such a treatment would be outside the scope of a single contribution.

In the following sections, I shall look at the drafting process more carefully in order to see how it conforms to or diverges from the criteria of sustainable forest management, specifically integration and biodiversity protection. For these purposes, I have used official preparatory documents, comments of various stakeholders, draft codes and newspaper articles as research material. I was also able to use interviews conducted by Maria Tysiaschniouk in the Russian Duma,[6] environmental NGOs and ministries in 2004.

---

5   Intergovernmental Panel on Forests (IPF) 1995–1997, Intergovernmental Forum on Forests (IFF) 1997–2000 and UN Forest Forum since 2000.

6   The State Duma is the lower chamber and the Federal Council is the upper chamber of the Russian Parliament.

Evaluation of Sustainability: Integration During the Drafting Process

Despite the huge forest resources of the Russian Federation, forestry does not play a significant role in the Russian economy. The low economic performance of the forest sector was the most important reason why the government of the Russian Federation wanted to have the latest forest law reform. The government wished to attract more investors into the forest sector and to make forestry more profitable in general (Government 2005).

The drafting of a new forest code started in 2002, when the government of the Russian Federation asked the Ministry of Natural Resources, as well as other executive authorities, to prepare a bill. The government, however, was dissatisfied with the relatively conservative proposals prepared by the Ministry and preparation of the bill was transferred to the more liberal Ministry of Economic Development and Trade. The power struggle between the two ministries was visible throughout the drafting process. All in all, some thirty drafts were published by the two ministries – both together and separately – and the president's chamber published a proposal of its own.

In April 2005, the State Duma had the first reading of the draft. Hundreds of amendments were proposed. The second reading was postponed several times and it was finally held in November 2006. After that the new Forest Code was accepted without delay also in the third reading in the Duma and subsequently in the Federal Council. The president of the Russian Federation signed the new code into law in December 2006 and for the most part the new law came into effect already in January 2007. The drafting process was a long one, but not especially long for an extensive law reform.

Despite the international activities of the Russian Federation, it was apparent in interviews and the State Duma proceedings that the drafters of the Forest Code were not concerned about the environmental and social impacts of forestry as such, but rather about a lack of a positive economic impact due to, among other things, insufficient wood-processing capacity and large-scale import of raw wood. The drafters seemed to be quite honest about their objectives concerning the new forest code: the economy comes first and ecology second (e.g. Interviews 2004; State Duma 2005). The discussion of a new forest code centred on reorganizing and rationalizing forest administration and liberalizing the system of forest leases. Privatization of state-owned forests was also much discussed, but in the end was rejected.

According to the preparatory work on the new forest code, the integration principle was not yet on the political agenda of the Forest Code drafters. On the contrary, the Vice Minister of the Ministry of Economic Development and Trade, representing the drafters of the bill, stated during the first hearing that the protection of endangered species has nothing to do with enacting a new forest code (State Duma 2005). Similarly, the vice-chairperson of the Committee on Natural Resources did not consider co-operation with the Environmental Committee important, since the Committee on Natural Resources thought that forest legislation is not directly related to ecological issues (Interviews 2004). These two comments clearly indicate the separateness of ecological and forest issues in the Russian drafters' thinking. Certainly, some members of the Duma and the Environmental Committee of the Duma think in a more integration-friendly manner than the main drafters of the Forest Code, but they seem to remain a minority (Environmental Committee 2005; State Duma 2005).

Civil society alleged the drafters to be reluctant to discuss and take into account the opinions of civil society. This was not the only problem concerning the discussion of the new law. The drafting process was characterized by 'chop and change' law-making. Opinions about important questions of ownership and leasing, for example, changed back and forth in consecutive drafts. The bills did not seem to evolve and improve; rather, every new draft was a new surprise by which the drafters endeavoured to win the approval of the government and the president. This probably made the discussion and any integration attempts even more difficult, since the discussion constantly billowed around ownership, leasing and organizational issues.

The relatively short explanatory note (poyasnitel'naya zapiska) that the government of the Russian Federation gave on 2 February 2005 to the State Duma along with the draft code was very modest where environmental goals (Government of the Russian Federation 2005) were concerned.[7] The main parts of the text were devoted to the effectiveness of forestry and forest administration, while environmental goals play a minor role; they are only mentioned very generally when briefly listing the overall ecological, economic and social goals of the draft code. The two ecological objectives mentioned in the explanatory note are protecting the sustainable state of the forest reserves and improving the quality of forests to fulfil their ecological functions. Both goals are ambiguous, leaving a lot of room for interpretation. For example, sustainability in the narrowest sense can mean that the annual amount of logging does not exceed the annual growth of forests. This would leave out most ecological and social aspects of sustainability.

The lack of environmental consideration and concrete environmental requirements in the draft code were strongly criticized by, for example, Russian ENGOs and the Environmental Committee of the State Duma. ENGOs were worried, for instance, about the fate of Group I forests, which include different kinds of protection zones and areas, and they made their own suggestions to improve the draft code (Kommentarii 2005). Even the Industrial Chamber of Commerce expressed its concern about the weakening of the Group I forests (Torgovo 2005). The Environmental Committee of the State Duma gave its own statement about the draft code in April 2005. The Committee drew attention, for instance, to the lack of

---

7 A serious shortcoming in Russian legislative work is the lack of clear rules for preparing the explanatory notes that the government gives to the Duma together with a draft code (OECD 2005, 13). One task of drafters is to provide legislators with background information and expert knowledge, which in turn contribute to the implementation of the new law (Tala 2005, 134). A clear and transparent motivation for proposed changes makes discussion easier. The relatively short explanatory note did not fulfil these criteria.

stated objectives for the draft code and to the fact that the protection of biodiversity was only mentioned in the list of principles (art. 1), but not operationalized in other provisions of the draft code (Environmental Committee 2005). These proposals were not discussed in the Duma. The Duma mainly discussed the forms of ownership and lease and the competence of the regions. The microphone was sometimes switched off if a Duma member started to criticize the drafting process rather than merely ask about the law draft or commenting on it (State Duma 2005). This happened, for example, to a representative of the Environmental Committee when he was about to describe the opinions of ecologists and foresters that the Environmental Committee had received regarding the draft code.

Integration and the New Forest Code

It seems that the international discourse on SFM and environmental agreements have hardly changed anything in the official thinking about forest management in the past decade. The Forest Code has remained more or less the same as regards ecological sustainability. The main changes have taken place in forest administration and in the rights and duties of leaseholders. The long-lived administrative bodies, leskhozes, were discontinued, official surveillance of forest use was streamlined by, for example, replacing licences by annual forest declarations, and forest management tasks were mostly left to leaseholders. Thus, the general tendency was towards liberalization of forest legislation. Choosing auctions as the only form of selecting leaseholders reinforces this tendency: it is no longer easy to favour local enterprises or set conditions that enterprises should employ local workers or contribute money or equipment to leskhozes, as used to happen when the old Forest Code was in place (Lehtinen 2006, 48–51). The one who makes the highest bid is now assumed to win the lease contract.

Environmental regulation partly worsened as a result of the Forest Code reform: e.g. the requirement of ecological assessment of forest plans was removed from the Law on State Ecological Assessment (Art. 16 of the Law on the Implementation of the Forest Code).[8] The Law on the Implementation of the Forest Code changed also some parts of the Law on Nature Conservation Areas. Construction in certain parts of national parks is no longer prohibited, but restricted. For example, the government can grant permits for the construction of sport facilities and similar projects.

---

8   Art. 89.1 of the Forest Code stipulates that 'any forest development plan shall be subject to state or municipal review following procedures established by the authorized federal executive body'. The wording is vague and it is not clear whether 'state or municipal review' is comparable to the previous requirement of an ecological impact assessment. Also the 'Rule on carrying out a governmental or municipal assessment of a forest development plan' (14.5.2007, no. 125) does not stipulate anything specific about assessment of environmental or ecological impacts of a forest plan. The plan should simply be in accordance with the federal legislation and a regional forest plan.

Construction is in fact promoted in the new Forest Code. Constructing buildings in forest areas leased for hunting, recreation, farming and collecting non-timber forest products was made much easier than before, which can have adverse effects especially near densely populated areas. In the worst case, the unclear status of 'special protection parcels of forests'[9] makes it possible to build hunting huts, recreational cottages and other structures on these protection parcels. Article 41 (forest use for recreational activities) requires that natural landscapes, wildlife, plants and water bodies must be preserved within the forest parcels made available for recreational activities, but this does not necessarily mean prohibition of construction. On the other hand, article 102.5 stipulates that activities which are incompatible with the purpose of their designation – which could be, for example, water protection, protection against erosion or location of the habitat of an endangered animal – are prohibited in protection forests and in 'special protection parcels of forests'. Yet, it would be better to specify an authority that permits other activities than permissible logging in these special forests. It is unclear whether the leaseholders are now allowed to use their own discretion in determining which activities are acceptable.

There were also some changes related to water protection zones, which are mentioned in both the Forest Code and the Water Code; the latter was revised in tandem with the Forest Code. The old governmental decree on water protection zones (23.11.2996, no. 1404) ordered water protection zones up to 500 metres along rivers – depending on the length of the river – and 300–500 m around lakes. According to article 65 of the new Water Code (3.6.2007, no. 74-FZ) the water protection zones along rivers encompass a maximum of 200 metres and around lakes 50 metres. Clear-cutting of forest is forbidden in water protection zones, but other forms of cutting are possible in accordance with the Forest Code (Water Code art. 63.2).

Even if integration of environmental aspects was lacking in the drafting process, it does not mean that they were totally lacking in the drafts of the Forest Code. Many long-established considerations, such as the categorization of forests and different kinds of protective zones around water systems, railways, main roads and industrial areas, remained in the drafts and stand almost unchanged in the approved version of the new Forest Code.[10] However, the possibility of creating a protection zone around spawning sites of valuable fish species was removed in the case of Group I forests (art. 102.2).[11]

---

9   According to article 102.3 of the Forest Code special protection parcels are, for instance, the habitats of rare and endangered species, forests with relict and endemic plants and forest edges adjacent to forestless areas. These parcels *may* be established in protection and production forests (art. 102.4).

10   The categorization was changed. The old category of Group II forests (e.g. forests of densely populated areas and in areas with few forests) was removed and the new groups are: protection, production and reserve forests. The fate of old Group II forests is still somewhat unclear.

11   The term 'pre-tundra forests' was changed to 'forest tundra zones' (art. 102.4). It remains to be seen what implications this will have for the protection of tundra forests.

The protection zones of Group I forests are clearly positive aspects of the Forest Code as such and prove that integration of environmental issues has already taken place to a certain extent. There are, however, certain shortcomings in forest regulation that decrease the positive effect of integrated rules, such as ambiguous wording, contradictory regulations and the respective powers of different authorities being unclear. The drafters seemed to want to reject further integration and leave many questions concerning, for example, endangered species and integrating nature protection in forest planning out of the current reform. Regional forest plans could have been developed in the direction of landscape ecological planning, for example, but it seems that they are still very technical documents with age structures and amounts of allowable loggings. It was not recognized during the drafting process that ecological issues are an integral part of forest issues and that forest management and biodiversity issues should be considered simultaneously. It is, of course, quite appropriate to regulate matters such as species protection during forest operations in laws other than the Forest Code. Russian legislation, however, contains no clear rules to this effect. There is, for example, a provision in the Law on Environmental Protection stating that the so called Red Book species are protected and it is forbidden to deteriorate their habitats and that more detailed rules will be given in the environmental legislation. However, there is no effective legislation that would properly implement that provision (see Pappila 2005).

There was no effort to strengthen the means of biodiversity protection that already existed in the legislation or to introduce new instruments of SFM. One way to implement the principle of integration could have been to improve the co-operation of the environmental and forest administrations in making forest management plans, to link the Forest Code to clear norms on the protection of endangered species and to make well-defined and transparent regulations about which protection and protective forests cannot be used for building or other activities. Introducing public participation in the forest leasing process, as the Environmental Committee of the State Duma proposed, would have been another way to integrate sustainable development in legislation. Leased forest areas can be huge, e.g. two million hectares, and a single rental agreement may thus affect the lives of thousands of villagers, whose livelihood is often partly dependent on berries, mushroom and other non-timber products found in forests.

Environmental aspects were integrated at a certain level into Russian forest legislation even before the latest law reform, as mentioned above, but the integration is not yet coherent, and SFM is not considered as a process that should be taken into account whenever laws are being amended. The reasons for neglecting the integration of environmental aspects can be found partly in culture. Strengthening the Great Power ideology and the national economy at any cost goes hand in hand in Russia and diminishes the weight of environmental policy (Tynkkynen 2005, 291). In addition, the idea of boundless forests and the development-positive standpoint persistently prevail in Russia (Tynkkynen and Massa, 2001, 14–15). Biodiversity and other environmental issues related to forests are not considered urgent, because the Russian Taiga is considered endless and there are other, more visible problems, e.g. widespread illegal logging and excessive forest fires.

It could thus be concluded that the integration principle as a part of sustainable development and sustainable forest management has not affected the attitudes of the federal government. In addition to ignoring the integration principle, the government has not sought new regulative instruments to enhance either ecologically or socially sustainable forest management.

Integration Since the Forest Code

Full implementation of the new Forest Code requires an array of decrees and orders. Many of them were already accepted during 2007. I have gone through some of them to see whether the integration principle has been taken into account in them.

At first glance the view was unpromising: in July 2007 the Ministry of Justice refused to register the new Decree on Logging Rules, because it contradicted the Forest Code.[12] The decree, drafted by the Ministry of Natural Resources, would have allowed clear cutting in special protection parcels of forests (WWF 2007). Article 107.2 of the Forest Code prohibits clear cutting in special protection parcels of forests, except for some special cases (article 17.4). The positive aspect here was that the Ministry of Justice did return the decree for further processing.

In spite of the difficulties with this one decree, there have been improvements in the new Logging Rules (Pravila Zagotovki Drevesiny, 16.7.2007 no. 184, Ministry of Natural Resources). According to article 13, it is now possible – although not required – to leave valuable standing trees in a logging area in order to enhance biodiversity if this does not hinder forest renewal. Retention trees used to be considered harmless for forest health in previous logging regulations. Article 10 includes a general requirement to protect Red Data Book (i.e. endangered) species and their habitats during logging operations. It is a very general rule and does not clearly state how extensively the area of a habitat should be protected and whether loggers should make a special species inventory or take into account only easily recognisable or known habitats (e.g. a habitat protected as a 'special protection parcel of forests' according to article 102.3 of the Forest Code). Nevertheless, it is an improvement, since there was no clear requirement to 'protect habitats' in the old logging regulations, except for the regulation of sanitary logging.[13]

There are, however, also some features in the new Russian Logging Rules that could be considered as detrimental from an ecological point of view. According

---

12   The Logging Rules Decree is perhaps the most important one in implementing the logging provisions of the Forest Code. There are also other, more detailed rules for different forms of cutting and for different regions.

13   There is also a prohibition against cutting viable valuable tree species (e.g. oak, beech, cedar and ash) on the edge of their area of distribution (article 11). The rules also protect cedar forests (article 12).

to the Logging Rule, the maximum clear cutting area varies from 5 to 50 hectares depending on the forest type and geographical area (article 44.1).[14] However, for various reasons a clear cutting area may be extended 1.5 times i.e. to a maximum of 75 ha. Due to rather vague wording, 75 ha may become the norm for clear cuttings in many areas.

The Order on the Contents of the Forest Development Plan (6.4.2007, no. 77) by the Ministry of Natural Resources requires a leaseholder to report, among other things, characteristics of nature protection areas located on the lease area, as well as information about animals and water bodies and the forms and aims of species and water protection. It is a positive addition, but the problem is that these formulations are very general. It is unclear, for example, what 'animals' means in this case; it is not a reference to endangered species mentioned in the Red Data Books. Nevertheless, the integration of biodiversity protection seems to work better on the implementation level than in the Forest Code.

Ironically, almost one year after adopting the new Forest Code, the government of the Russian Federation published 'The Plan of Growth of Forestry in the Russian Federation until year 2010', in which the it states that the system of biodiversity protection in production forests should be developed. This issue was rejected during the Forest Code reform. Then again, the government's idea to develop biodiversity protection is not included in the implementation strategy for the Plan of growth (Government 2007).

The ecological sum of the Forest Code reform is somewhat contradictory. The Forest Code itself clearly lead to a weakening of biodiversity protection. On the other hand, in the new logging rules the Ministry of Natural Resources has introduced some new regulation to enhance biodiversity in production forests. Moreover, the decision of the Federal Forest Agency to form 31 new model forests in Russia is a step towards sustainable forest management (Federal Forestry Agency 2007).[15] Discussion of ecological issues was not considered relevant by either the Russian government or the majority of the Duma. Taking into consideration the proposal to

---

14   For example, in North-West Russia, the Urals and Siberia, the maximum clear cutting area for pine and spruce forests are 40–50 ha. Smaller maximum areas, 5–10 ha are mainly for southern parts of Russia.

15   'A model forest is a place where the best sustainable forest management practices are developed, tested and shared across the country. It is an ideal laboratory for conducting research on sustainable forest management.' http://www.modelforest.net/cmfn/en/about/whatis.aspx (20.11.2007) International Model Forest Network describes model forests as follows: "Each model forest is established as a working-scale model aimed at effecting a transition from conventional forest management to management for sustainable forest production and environmental conservation. Each model forest attempts to demonstrate sustainable and integrated forest management, to transfer the knowledge to forest managers, and to apply technological advances when appropriate. Each model forest is managed through a partnership of stakeholders in the area and demonstrates the integrated management of key resources through ecologically sound forestry practices." http://www.imfn.net/en/ev-22934-201-1-DO_TOPIC.html (20.11.2007)

revise the protection of biodiversity in production forests, there is still a possibility of more positive changes in Russian forest legislation. There is a risk, of course, that the turbulent transition period from the old system of the 1997 Forest Code to a new one will undermine the minor positive changes that were mentioned above. Simultaneous change of the whole regional level forest administration, changes in the powers of the regions and the federal authorities and in the rights and duties of leaseholders may lead to uncontrolled logging with numerous uncertainties regarding the rights and duties of each stakeholder. The first experiences of the new code can be described as a fair degree of bewilderment and discontent and the transition period has been considered too short.[16]

A Forest Policy: A Cure for All Evils?

The anti-environmental attitude of the federal government and the insufficient level of integration of environmental aspects into forest legislation are clearly obstacles to improving the ecological sustainability of Russian forest governance. Both could also be mentioned as reasons for the lack of a coherent forest policy (programme) in the Russian Federation – in addition to ongoing economic, political and social processes of transition (Torniainen et al. 2006). Looking at the question from another point of view, one could claim that the process of creating a real forest policy for Russia could improve the level of sustainability. The Intergovernmental Panel on Forests (IPF) has emphasized certain aspects in developing and implementing a national forest policy or programme, including appropriate participatory mechanisms to involve all interested parties, decentralization, holistic and intersectoral approaches and empowerment of local and regional governments (UN 1997, 6).

Firstly, following the proposals of the IPF, development of a national forest policy could give the public a chance to take part in determining the direction of the development of the Russian forest sector (Humphreys 2004, 208; Reeb 2004, 88). Secondly, if the forest policy were a real, coherent and extensive one taking the integration principle seriously, it would help to integrate ecological sustainability (e.g. the spirit of international environmental agreements and declarations) into forest legislation and other forms of forest governance. Thirdly, such a policy might help to moderate the federal government's very heavy focus on the economy in forest issues. Fourthly, it could help to make permanent decisions on the centre/regions and other organizational questions. In addition to ad hoc changes in the forest legislation (e.g. in 2004 the maximum leasing period was prolonged from 49 to 99 years but in the new Forest Code it was again shortened to 49 years), constant changes in the powers of different authorities and the competent authorities make

---

16   This was the message of the most speakers in the international seminar 'Modern forest legislation of the Russian Federation: First practical results' in St. Petersburg State Forest Technical Academy in November 2007.

it very difficult to pursue any sort of forest policy, except for a policy of everyday survival. Teplyakov et al. (1998, 16) remark that between 1917 and 1992 Russian/ Soviet central forest agencies were reorganized 20 times and completely liquidated three times. This means that agencies were changed at least every four years. The same trend of constant administrative changes has continued since 1992. The new Forest Code of 2006 has led to another series of administrative changes, e.g. the liquidation of leskhozes.

Forest policy is still a relatively new concept in Russia, since it was not considered important in the Soviet Union. There is not yet any coherent forest policy in Russia today. But again, even a lack of forest policy is a policy of sorts. The current Russian 'no policy' forest policy consists of forest legislation, target programs and strategy documents, which are contradictory, inconsistent and changing constantly (Saastamoinen and Torniainen 2005, 29; Lehtinen 2006). There are, for example, the Reforestation Programme and the Federal Target Programme forests of Russia to enhance technical capacity building (Forest News 2005b) and the renewed Concept of Growth of Forestry in the Russian Federation until year 2010 (Government 2007). However, this has not been a solid basis for a profound reform of the forest legislation, administration and forest sector as a whole. Law drafting problems may well have been partly due to the lack of discussion of overall objectives and coherent background work. The lack of forest policy was also mentioned as a reason for law-making difficulties in the discussions of the Duma (Duma 2005). For example, the more forestry-oriented Ministry of Natural Resources engaged in a tug of war with the more (economically) liberal and industry oriented Ministry of Economic Development and Trade and the lack of national forest policy probably aggravated the quarrel. The need for a new forest policy has been widely acknowledged among Russian scientists and politicians (Veijola 2004), but their political influence has apparently not been strong enough.

Saastamoinen and Torniainen (2005, 31) take the view that, in a way, national forest policy has been formulated at the same time as the new forest code has been drafted. This is certainly true; important definitions of policy have been made in the new Forest Code. Then again, this way of proceeding does not lead to a consistent and transparent forest policy formation and has apparently prolonged the drafting process. It is not a genuine forest policy as the concept is currently understood: it should take into account, for example, the requirements and general principles of international environmental agreements. The public (e.g. indigenous people and NGOs included) should be able to participate in formulating the policy, and different aspects of sustainable forest management should be integrated into it. Now there is a great risk that after enacting a new Forest Code at the end of 2006, there will be no room or interest for a profound discussion of a national forest policy, since many important aspects of forest governance have been nailed down in the new Forest Code. In all likelihood, there will be numerous piecemeal changes in the new Forest Code in the coming years. Nevertheless, this is not likely to change the situation; after the great effort to enact the new Forest Code, there will be no big forest policy discussion in the near future in Russia.

The lack of a coherent national policy (programme) leaves more room for other stakeholders to formulate Russian forest policy de facto. Torniainen et al. (2006) argue that industry will become a more important player in the development of Russian forest policy, because of the privatization of many forest management tasks. Alongside this tendency towards liberal, market-driven forest governance, the rule of law is still weak in Russia; official institutions often do not work as they should (see e.g. Rose and Munro 2002; Nysten-Haarala 2005). This diminishes the effectiveness of laws and gives more room to all kinds of unofficial institutions and rules. As an example of this, some Russian entrepreneurs consider voluntary FSC forest certification more successful in regulating forestry than forest legislation. Market-based economic sanctions e.g. loss of income and cross-border business relations are considered more effectual than occasional administrative inspections and fines (Tysiaschniouk and Reisman 2005, 157–8).

Thus, if the chaotic state policy does not lead to a functioning forest policy, other stakeholders will fill the gap in one way or another. This appears to be a traditional case of state failure, but alongside it one can recognize a trend of decentralized regulation and the network of global forest governance, which in turn are partly due to international state failure. Nation states have failed to agree on effective international means to reduce the negative externalities of forestry and global trade in forest products. Increasing voluntary forest certification and protection activities of industry in co-ordination with ENGOs illustrate the tendency of transnational forest governance (see also Chapters 7 and 9 in this volume). International and national state failure has led in many countries to civil society initiatives to protect old-growth forests and pursue sustainable forest management. It is not only a question of states being unable to govern forest issues efficiently, but a failure to govern forests ecologically and in a socially sustainable manner.

Even if there apparently exists a de facto Russian forest policy, i.e. a combination of fluctuating state governance and the operations of Russian and transnational enterprises and NGOs, it does not fully make up for a national state-run forest policy as a potential basis for sustainable forest management. A widely debated national forest policy could also help to elucidate the connections between the Forest Code and other environmental legislation and contribute to the integration of environmental considerations into forest legislation. Nevertheless, without enough political will and commitment to the principles of sustainable development and to a national forest policy itself, a policy document would not cure anything. Law reform is certainly not enough to ensure stable or sustainable development of the forest sector of a country where the economic sector, legislation and administration remain in a state of continuous change and rapid transformation. The establishment of a new Department of Forest and Water Policy in the Ministry of Natural Resources in May 2006 (MNR 2006) could be a sign of new appreciation of policy-making – or it could be another sporadic and temporary change in the forest administration.

Conclusions

The main aim of this chapter was to evaluate ecological sustainability as reflected in Russian forest regulation by assessing the law-drafting process and the new Forest Code. Different economic, political and social forces pull the development of the Russian forest sector in different directions. The attitude of the Russian government towards environmental protection and sustainable forestry did not strengthen the ecological sustainability of the Russian Forest Code in a way that international commitments would require. Utilization of natural resources is clearly a leading principle, and ecological sustainability is on the losing side. The rationale behind the government's attitude can be found in the 'Great Power with endless forests' thinking and in the general downplaying of environmental policy in Russia (Tynkkynen 2005).

The ecological sum of the Forest Code reform is somewhat contradictory. The Forest Code itself lead to weakening of biodiversity protection. On the other hand, the Ministry of Natural Resources has introduced certain new protection ideas in its implementing rules. Taking into consideration the governmental proposal to revise the protection of biodiversity in production forests (Government 2007), positive changes may be forthcoming in Russian forest legislation. Dividing forests into three categories and different kinds of protective zones forms a good basis for biodiversity protection and sustainable forest management.

Yet, there is a risk that the turbulent transition period from the old system of the Forest Code to a new one will undermine the few positive changes and even the old protection instruments that were mentioned above. Simultaneous modification of the whole regional level forest administration, changes in the powers of the regions and the federal authorities and in the rights and duties of leaseholders may lead to un(der)controlled logging and uncertainty about the rights and duties of each stakeholder.

The liberalization of the Russian governmental forest regulation can be seen as part of global economic liberalization, even though there are still many signs of protectionism and state paternalism in Russia. According to the most generally accepted idea of sustainable development – some call it weak sustainable development – the fact that economic growth or liberal theory underlies current global economic liberalization is not per se contrary to sustainable development (Baker et al. 1997, 13; de-Shalit 2000, 64). However, if economic liberalization is put into practice without due concern for ecological and social considerations, the resulting economic regime is likely to hinder sustainable development. This is more likely if society is not otherwise liberal, i.e. open for discussions and tolerant of a variety of opinions and ways of life. In this regard, the one-sided liberalization of Russian forest legislation, i.e. liberalization of forest leasers' rights only, may hinder sustainable forest management. It is clear that some aspects that could counterbalance economic liberalization are still weak in Russia: the role of environmental considerations in governmental regulation and the role of civil society. Then again, this liberalization could also open more opportunities for co-operation between enterprises and ENGOs, but potentially increasing co-operation should not be left without legislative support. The improvement of the protection of forest biodiversity should not depend on the uncertain success of co-operation among civil society stakeholders and the insufficient supervisory resources of ENGOs.

In addition to the lack of environmental ambition of the Russian government, another underlying trend that inhibits achieving sustainable forestry is the constant change in forest administration and legislation. In the case of Russian forestry, this not only concerns the transition period which started after the collapse of the Soviet Union. It has rather become a tradition and a supposed way of improving the Russian forest sector. A coherent and long-term national forest policy could be a way of getting off this path, and could improve the integration of sustainable development into Russian forest governance.

References

Adamowicz, W. and Burton, P. (2003), 'Sustainability and Sustainable Forest Management', pp. 41–46 in Burton et al. (eds).

Baker, S., Kousis, M., Richardson, D. and Young, S. (1997), 'Introduction. The Theory and Practice of Sustainable Development in EU Perspective', pp. 1–42 in Baker et al. (eds).

Baker, S., Kousis, M., Richardson, D. and Young, S. (eds) (1997), *The Politics of Sustainable Development. Theory, Policy and Practice Within European Union* (London/New York: Routledge).

Burton, P.J., Messier, C., Smith, D.W. and Adamowicz, W.L. (eds) (2003), *Towards Sustainable Management of Boreal Forest* (Ottawa: NRC Research Press).

Buttoud, G., Sohlberg, B., Tikkanen, I. and Pajari, B. (eds) (2004), *The Evaluation of Forest Policies and Programmes* (Saarijärvi: European Forest Insitute).

de-Shalit, A. (2000), *The Environment, Between Theory and Practice* (Oxford: Oxford University Press).

Duinker, P. and Trevisan, L. (2003), 'Adaptive Management, Progress and Prospects for Canadian Forests', pp. 857–892 in Burton et al. (eds).

Environmental Committee of the State Duma (2005), *Zaklyuchenie Gosudarstvennoy Dumy po ekologii na proekt federal'nogo zakona no. 136515-4* (The statement of Environmental Committee of the State Duma on Law Proposal No. 136515-4) (In Russian).

Glowka, L., Burhenne-Guilmin, F., Synge, H., McNeely, J.A. and Gündling, L. (1994), *A Guide to the Convention on Biological Diversity* (Geneva: IUCN, The World Conservation Union).

Government of the Russian Federation (2007), Rasporyazhenie Pravitel'stvo Rossiyskoy Federatsii ot 28 centyabrya 2007g. No. 1305-r., Kontseptsiya razvitiya lesnogo hozyaystva Rossiyskoy Federatsii do 2010 goda (Order of the Government of the Russian Federation of 28 September 2007. No. 1305-r., Plan of the Growth of Forestry in the Russian Federation until the year 2010) (In Russian).

Humphreys, D. (2004), 'National Forest Programmes as Policy Vehicles for Sustainable Forest Management, Findings from a Major European Research Project', pp. 207–216 in Buttoud et al. (eds).

Interviews (2004), The interviews include about 40 interviews conducted by Maria Tysiaschniouk among Russian members of the State Duma and their assistants, NGO workers and officials of the Ministry of Natural Resources conducted as part of a Finnish-Russian research project 'Governance of Renewable Natural Resources in Northwest Russia' funded by the Academy of Finland.

Kokko, K. (2003), *Biodiversitetttiä turvaavat oikeudelliset periaatteet ja mekanismit* (*Legal Principles and Mechanisms for Safeguarding Biodiversity*). (In Finnish.) (Helsinki: Suomalainen Lakimiesyhdistys).

Kortelainen, J. and Kotilainen, J. (eds) (2002), *Environmental Transformations in the Russian Forest Industry. Key Actors and Local Developments* (Joensuu: University of Joensuu).

Liberatore, A. (1997), 'The integration of sustainable development objectives into EU policy-making', pp. 107–126 in Baker et al. (eds).

MCPFE (1993), Resolution H1, General Guidelines for the Sustainable Management of Forests in Europe. Second Ministerial Conference on the Protection of Forests in Europe. 16–17 June 1993, Helsinki/Finland.

Nystén-Haarala, S. (2005), 'Lainsäädäntö ja taistelu luonnonvarojen hallinnasta Venäjällä', (Legislation and fight over the governance of natural resources in Russia). (In Finnish.) *Alue ja ympäistö* 34:2, 36–47.

Pisarenko, A.I. and Strakhov V.V. (1996), 'Socio-Economic Assessment of the Russian Boreal Fores'. Working Paper-96-58 (Laxenburg: IIASA).

Reeb, D. (2004), 'Towards Development of Criteria and Indicators to Assess the Level of Participation in NFP Process', in Buttoud, G. et al. (eds), pp. 87–102.

Rose, R. and Munro, N. (2002), *Elections without Order, Russia's Challenge to Vladimir Putin* (Cambridge: Cambridge University Press).

Rosenholm, A. and Autio-Sarasmo, S. (eds) (2005), *Understanding Russian Nature. Representations, Values and Concepts* (Saarijärvi: Gummerus)

Saastamoinen, O. and Torniainen, T. (2005), 'Venäjän metsäpolitiikan murros jatkuu (Transistional period of Russian forest policy is continuing). (In Finnish.) *Ostiensis* 1/2005, 29–31.

Schelhas, J. (2003), 'New Trends in Forest Policy and Management, an Emerging Postmodern Apporoach?', pp. 17–28 in Teeter et al (eds).

Stepanitski, V.B. (1997), *Kommentariy k Federal'nomu zakonu Rossiyskoy Federatsii ob osobo ohranyayemyh prirodnyh territoriyah* (*Commentary on Federal law of the Russian Federation, about specially protected areas*) (Moscow: Tsentr ohrany dikoy prirody).

Tala, Jyrki (2005), *Lakien laadinta ja vaikutukset* (*Legislative work and the effects of legislation*) (In Finnish.) (Helsinki: Edita).

Teeter, L., Cashore. B. and Zhang, D. (eds) (2003), *Forest Policy for Private Forestry. Global and Regional Challenges* (Wallingford, Oxon, UK : CABI Publishing).

Torniainen, T., Saastamoinen, O. and Petrov, A. (2006), 'Russian forest policy in the turmoil of the changing balance of power', *Forest Policy and Economics* 9:4, 403–16.

Tynkkynen, N. (2005), 'Russia, a Great Ecological Power? On Russian attitudes to environmental politics at home and abroad', pp. 277–296 in Rosenholm and Autio-Sarasmo (eds).

Tynkkynen, V.-P. and Massa, I. (2001), 'Introduction', pp. 11–26 in Tynkkynen and Massa (eds).

Tynkkynen, V.-P. and Massa, I. (eds) (2001), *The Struggle for Russian Environmental Policy* (Saarijärvi: Aleksanteri Institute).

Tysiaschniouk, M. and Reisman, J. (2002), 'Transnational Environmental Organizations and the Russian Forest Sector', pp. 56–71 in Kortelainen and Kotilainen (eds).

WCED (World Commission on Environment and Development) (1987), *Our Common Future, Report of the World Commission on Environment and Development* (Oxford: Oxford University Press).

Weiner, D. (2005), 'The Genealogy of the Soviet and Post-Soviet Landscape of Risk', pp. 209–236 in Rosenholm and Autio-Sarasmo (eds).

Yanitsky, O. (2001), 'Risk Society and Environmental Policy in Russia', pp. 27–48 in Tynkkynen and Massa (eds).

Yaroshenko, A.Y., Potapov, P.V. and Turubanova, S.A. (2001), *The Last Intact Forest Landscapes of Northern European Russia* (Greenpeace Russia and Global Forest Watch).

Internet-Based References

Federal Forestry Agency (2007), V Rossii budet sozdan 31 modelniy les (There will be 31 model forests established in Russia). Press release 1 November 2007. <http://www.rosleskhoz.gov.ru/media/news/16>, accessed 20 November 2007.

Forest News (2005b), Keynote Address by Valery P. Roshchupkin, Chief, Federal Forestry Agency at VII International Forestry Forum, St.Petersburg, Russia, October 4, 2005, "Russian forest sector performance in 2004–2005. Forest policy priorities at the national and international levels." <http://www.forest. ru/eng/news/index.html?AA_SL_Session=2d4928ade127cd7a4412e8aebfa74 d34&x=6499>, accessed 13 April 2006.

Government of the Russian Federation (2005), Poyasnitel'naya zapiska k proyektu Lesnogo Kodeksa Rossiyskoy Federatsii (Explanatory note on the draft of the Forest Code of the Russian Federation.) <http://www.forestlaw.ru/official_ texts/2/>, accessed 1 June 2006.

Kommentariy. (2005), Kommentariy rossiyskih nepravitel'stvennyh organizatsiy k proyektu Lesnogo Kodeksa RF (versiy, vnesennoy v Gosudarstvennuyu Dumu) (Comments of Russian environmental non-governmental organizations on the draft Forest Code of the Russian Federation (version that was given to the

State Duma). (In Russian.) <http://www.forestlaw.ru/opinions_comments/1/>, accessed 15 June 2006.

Lehtinen, L. (2006), Venäjän metsäoikeudesta (On Russian forest law). (In Finnish.) <http://www.idanmetsatieto.info/fi/cfmldocs/document.cfm?doc=show&doc_id=928>, accessed 15 October 2006.

MNR (Ministry of Natural Resources) (2006), Vneseny izmeneniya v strukturu MPR Rossii (Changes made in the structure of the Russian Ministry of Natural Resources.) (In Russian.) Press release 11 May 2006. <http//www.mnr.gov.ru/>, accessed 23 May 2006.

OECD (2005), Government Capacity to Assure High Quality Regulation. Regulatory Reform in Russia. OECD Reviews of Regulatory Reform. <http://www.oecd.org/dataoecd/14/45/34985412.pdf>, accessed 7 June 2006.

Pappila, M. (2005), Venäjän metsälainsäädäntö ja kansainväliset sitoumukset biologisen monimuotoisuuden turvaamiseksi. (Russian forest legislation and international commitmets to protect biodiversity.) (In Finnish.) <http://www.edilex.fi>, accessed 2 June 2008.

Russia's Report (2003), Russia's Report on the Montreal Process; Criteria and Indicators for the Conservation and Sustainable Management of Temperate and Boreal Forests, 2003. <http://www.mpci.org/rep-pub/2003/RussiaE/main.html>, accessed 11 July 2005.

State Duma (2005), The proceedings of the session of the State Duma of the Russian Federation on the 22nd of April 2005. (In Russian.) <http://wbase.duma.gov.ru/steno/nph-sdb.exe?B0CW[F11&20.04.2005&F11&27.04.2005&F11&&F258&^&]H2082>, accessed 27 April 2006.

Teplyakov, V.K., Kuzmichev, Ye. P., Baumgartner, D. M. and Everett, R. L. (1998), A History of Russian Forestry and its Leaders. <http://www.forest.ru/eng/publications/history/>, accessed 19 May 2006.

Torgovo Promyzhlennoy Palaty (2005), Rezhenye Komiteta Torgovo Promyzhlennoy Palaty po Lesnomu Kodeksu RF (The decision of the Industrial Chamber of Industrial Commerce about the Forest Code of the Russian Federation) (In Russian.) <http://www.forestlaw.ru/opinions_comments/2/>, accessed 15 June 2006.

UN (1997), Report on the Ad Hoc Intergovernmental Panel on Forests on its Fourth Session. Commission on Sustainable development, Fifth Session, 7–25 April 1997. <http://daccessdds.un.org/doc/UNDOC/GEN/N97/069/11/PDF/N9706911.pdf?OpenElement>, accessed 9 January 2007.

Veijola, P. (2004), Kansallinen metsäpolitiikka ja metsälaki. Muistio 2.6.2004, UM (National forest policy and the Forest Act. Memorandum 2 June 2004, Ministry of Foreign Affairs.) (In Finnish.) <http://www.idanmetsatieto.fi/>, accessed 10 January 2006.

WWF (2007), Minyust "zarubil" Pravila zagotovki drevesinu (Ministry of Justice "cut down" the decree on logging) (In Russian) 4 September 2007 <http,//www.wwf.ru/resources/news>, accessed 12 September 2007.

Chapter 5

Fishery Governance in Northwest Russia

Larissa Riabova and Lyudmila Ivanova

Introduction

Fishery has traditionally played a significant role in economies of many regions of the Russian North. In this chapter we focus on Northwest Russia, particularly the Murmansk Region. In this region, for generations the fish resources of the Barents and the White Seas provided a livelihood for both the indigenous people (the Kola Sami) and many small coastal communities of more recent settlers (Novgorodians and Pomors). Under the Soviet regime, the Murmansk Region was developed and strengthened by the growth of the industrial fisheries sector (Eikeland, Riabova and Ivanova 2005). Nowadays, about 14,000 persons are employed in fishery-related activities in the region. The fishing fleet consists of 69 coastal vessels and 75 sea-going trawlers. The fleet accounts for about 20 per cent of total Russian landings (Larsen 2004).

In the 1990s, Russia began to introduce a market-based regime for governing fishery. A new governance framework began to form. It was proclaimed that the new framework should be based on the principle of federalism. This implied significant changes first of all in the interaction between federal and regional levels of state power. From today's perspective, this process has been exceptionally complicated as compared to similar processes in other sectors of the Russian economy. It is still incomplete and in our view is far from completion. At the same time, there is a strong international development towards the central government losing its power to other actors – civil society, private companies and international actors. Some signs of this process, which can be described as the rise of multilevel governance, have become noticeable in Russia, too. However, it is not quite clear whether these signs can be viewed as constituting a rise of multilevel governance (or at least as pointing to the start of this process) or whether they are just fragmentary occurrences having nothing to do with the phenomenon.

We will analyze the development of the governance framework of fishery in Northwest Russia during the last two decades, applying the concept of multilevel governance. The main question to answer is whether this development is headed towards multilevel governance. We do not include the international perspective in our discussion but rather focus on national aspects of governance.

The chapter is based on several types of sources. The primary sources are qualitative interviews of the representatives of regional authorities, fishery professionals and scientific experts, representatives of fishers' associations, leaders and managers of fishing companies, leaders and members of fishery *kolkhozy*, residents of the fishery-based communities, and representatives of ecological NGOs in the Murmansk Region. The interviews were carried out in the period 2003–2006. We also draw on the contributions and reports of other researchers, as well as publications from regional and local newspapers. In addition, we use knowledge received through our participation in the project 'The Battle for the Resource Rent' in cooperation with Norut NIBR Finnmark (Alta, Norway), in 2002–2006.

The Russian Mode of Governance from the Perspective of Multilevel Governance

One of the most important developments in modern societies in recent decades has been the transformation of state-based governing mechanisms and the rise of new governance arrangements. Along with this process, the theoretical approach summed up as *governance* became very common. In the literature, governance has several meanings. Some authors consider that from the theoretical perspective governance should refer to the phenomenon whereby many public functions increasingly seem to be carried out by actors other than the classical government institutions of the nation-state and its subdivisions. From this point of view, governance consists of governing mechanisms which do not rest on recourse to the authority and sanctions of government (Stoker 1998). Public administration is thus increasingly becoming 'unbounded', involving various public, non-governmental and private actors in the process of making decisions on public goods. Another group of researchers has a broader understanding of governance as the action or manner of governing, or the processes by which common problems are dealt with, and differentiate between several modes of governance, for example, governance by government, governance by networks, or governance without government.

As a rule, new forms of governance rely less on the state as the institutional backbone and hierarchical centre of society. The authority of the state becomes divided or shared and public policy becomes the outcome of complex relations characterized by the term multilevel governance. '*Multilevel*' refers to a variety of forms of (private or public) decision making, authority, policy making, regulation, organization, ruling, steering, and so on; characteristic of these is a complex interweaving of actors operating at different levels of formal jurisdictional or administrative authority, ranging from the local level, via the regional and national level, to the regional and international or global level. The rise of multilevel governance transforms the capacity, location and performance of the traditional (state-centred) mechanisms of power, responsibility, legitimacy and accountability (Bache and Flinders 2004; Peters 1996; Rhodes 1997).

The rise of multilevel governance is recognizable. It is widely perceived that, among other phenomena, it is seen in the declining importance of the nation-state and national legal systems as the central locus of public authority and in international relations, the declining role of hierarchy in public and private decision making, the rise of open methods and architectures of coordination, the use of 'soft law' and the development of 'coordinated decentralization' in the governance of firms and industrial relations. It also appears generally in the increasing significance of networks and complex negotiating systems involving public authorities, (semi-)private actors, non-governmental organizations (NGOs), social movements, multinational corporations, interest groups and expert organizations (Bache and Flinders 2004; Pierre and Peters 2000).

In this chapter we apply the concept of multilevel governance to illuminate the development of the governance framework of fishery in Northwest Russia during the last two decades. To set up the context for this discussion, we begin with an analysis of the general situation in Russia in regard to the development of the national mode of governance during the last decade.

The Russian Constitution defines Russia as a federal state. During last two decades, transformation of the mode of national governance has been based on the principle of federalism. By definition, federalism is the negation of a strong centralized state. It is territorial democracy and in Russia, with its authoritarian and unitary past, the development of federalism is strongly connected with a significant change of relations between the centre and regions. In the Russian Federation – a multinational country with a complicated administrative-territorial structure – relations between the centre and regions have always been one of the most important policy factors exceeding the limits of the interactions between the two power levels and have been an indicator of the condition of the political system as a whole (Turovsky 2006). Since the declaration of Russian sovereignty the vector of these relations has changed its direction repeatedly. Starting from 1990, three periods can be identified in the interactions between the federal power and regions in the Russian Federation (Lapina and Chirikova 2004):

- 1990–mid-1993: spontaneous decentralization
- mid-1993–2000: asymmetric contractual federation
- 2000–present: policy of new centralism.

In the period 1990–mid-1993, relations between the centre and regions were destabilized by the disintegration of the USSR. The Russian state was too weak to establish universal institutional and political frameworks for the region-members of the Federation; the central power lacked the administrative resources for governing the huge Russian territories. To fill in the power vacuum, B. Yeltsin delegated broad authority to regions by concluding a political compromise with regional leaders in keeping with the formula 'power in exchange for political loyalty'. Personal, intra-elite agreements kept Russia from disintegrating but only became a temporary solution to the problems that had appeared due to spontaneous decentralization.

In the period from mid 1993 to 2000, an asymmetric contractual federation formed in Russia. After the Constitution of 1993 was adopted, legal decentralization of state power became possible. The new Constitution determined the political structure of the country as a federal state with a republican form of governance. The main features of the asymmetric contractual federation in this period were the bilateral nature of interactions between the state leader and regional leaders, inequality between regions and in relation to the centre and the lack of universal rules in centre-region relations. The word 'asymmetric' in the description of this period refers to these features, especially the last two. Bilateral agreements between the centre and the regions became a political trademark of this period. The political compromises reached in these years made it possible to partly stabilize relations between the federal power centre and regional elites.

The primitive federalism of this period had a pronounced anti-crisis orientation. Regions received extensive resources and powers – a development which Moscow did not like. This enabled regions to solve many of problems that the federal centre could not solve on its own. In spite of the imperfectness of the asymmetric federation, the centre and regions achieved a certain balance, and both power levels began to learn to coordinate their interests. Starting in the mid-1990s a new trend took shape – growing attempts by regional elites to gain positions of power of the federal centre. By the end of the 1990s 'rebellious' regional elites began their own preparation for the election cycle of 1999–2000. The prospect of losing power forced the elite of the federal centre to mobilize resources to ensure the victory of its candidates.

In 2000, the election of V. Putin as President marked the beginning of a new stage in federal policy. It meant the creation of a new governance system and a concentration of administrative, economic, political and other resources in hands of the federal centre. The basic argument in favour of the new centralization was effectiveness: from the point of view of federal power, decentralization reduced the effectiveness of state and municipal governance (Nechaev 2008).

The new policy was incompatible with the preservation of bilateral agreements between Moscow and the regions; regional leaders lost the status of federal politicians (as a result of the reform of the Federation Council, they left the upper chamber of the Parliament and left their senate seats to their appointees). Centralization spread into practically all of spheres of centre-regions relations. Regional legislation was brought into accord with federal, and a single legal space was restored in Russia. All over the country conditions ensuring free movement of goods and capital were created, securing a single economic space as well. Financial resources began to concentrate in the centre not least due to changes in tax legislation in favour of the federal level (Lapina and Chirikova 2004).

The new centralization touched administrative and political life. With 'vertical power' restored, the institution of 'federal intervention' was created, making it possible to remove universally elected regional leaders from their positions upon the decision of the federal centre. Federal districts were created to establish better control over the activities of regional branches of federal bodies. Political centralization included a reform of electoral legislation, the removal of regional parties from the election process and unification of the system by which representative power bodies are formed in regions of the Russian Federation. The rationale here was that the decision-making process would be simplified and would become more efficient, that 'centre-region' relations would be placed into a legal space and rationalized by submitting them to general norms and rules, that the president would openly communicate with regional leaders, mainly within the new collective body representing regional interests (the State Council) and that regional leaders would focus on intraregional issues and not demand participation in general federal policy.

In the first stage (2000–2002), the reform of federal relations – carried out 'from above' without consultations with the regions – destabilized centre-regions relations. The customary communication channels were broken and prominent positions in federal ministries were filled by people having no strong relations with the regions (Lapina and Chirikova 2004). By 2003, relations between the centre and regions stabilized. In the legal, administrative and economic spheres, many of the tasks set by the federal power were fulfilled; 'presidential vertical power' in regions was built up, but regional elites kept autonomous political resources. Putin's first four-year term in office resulted in a regime of controlled pluralism in which the federal power strengthened its positions but did not manage to free itself from the political influence of regional elites. In the new presidential term (from the beginning of 2004), the Kremlin had to make a choice – to either continue the policy of political coordination or complete the centre-oriented federal reform (ibid).

The choice was made in favour of centre-oriented reform. On September 13, 2004, in his speech at the session of the government of the Russian Federation, Putin formulated proposals that formally were directed to solving 'issues of the country's unity' but in fact changed the system of political power. Among the proposals were a refusal to permit universal elections of regional leaders and the introduction of indirect elections of regional governors by regional parliaments on the basis of candidates proposed by the president. Putin's proposals, most of which were carried out, called into question the future of Russian federalism and, in the estimation of some experts, created a basis for a return to a unitary state.

How can the processes described above be explained from the multilevel governance perspective? If we choose the narrow understanding of multilevel governance, whereby many public functions increasingly seem to be carried out by actors other than the classical government institutions of the nation-state and its subdivisions, an analysis of region-centre relations provides us with no evidence of development towards multilevel governance. This is because formally both the federal and regional levels in the Russian Federation represent state power and any redistribution of power between these two levels makes no difference, since power continues to belong to state bodies.

However, we suggest that the broader view of multilevel governance can be applied in the Russian case. We believe that in extremely centralized countries the process by which the centre loses its power to the subnational, that is, regional level, can be seen as a process constituting the development of multilevel government and that centre-region relations have to be included in the analysis. Thus, in this perspective, we wonder whether we might consider the period when the federal centre in Russia was losing its power to the regions in 1990–2000 as a rise of multilevel government in the country. Taking into account the extremely strong role of the centre throughout Russian history until the early 1990s as well as the fact that the regional level only became a real actor in governance quite recently – opposing the federal centre in many spheres in 1990–2000 and solving many tasks independently of the centre since 1990 – we are tempted to answer this question in the affirmative. Nevertheless, since 2000 another trend, a re-centralization of governance and a weakening of the regions' positions, has been observed and this definitely does not point to a rise of multilevel governance in the country.

An important indication of the rise of multilevel governance is the increasing involvement of the local level in the governance process. In the Russian Federation governance functions on the local level are carried out primarily by local self-governments. Local self-government is guaranteed by the Constitution and does not fall within the authority of the state. By signing and ratifying the European Charter of Local Self-Government, Russia confirmed the intentions declared in its Constitution. In the 1990s, the Russian Federation took several steps towards the formation of a system of local self-government – from the adoption of a number of federal and regional laws to the election of municipal councils and the election of mayors throughout the country. Today there is evidence of progress in the development of local self-government in the country. Most important have been the formation of a legal basis for democratic self-governance and an increase in people's involvement in the governance process on the local level.

Nevertheless the reform of local self-government in the Russian Federation has been only partly successful. Lately (since 2000) the most important characteristics of self-government – local democracy and local autonomy (i.e., the possibility for municipal power bodies to carry out their activities independently of control by higher governance bodies) – have undergone such changes that, in the very critical estimations of some authors, 'the present situation should be compared to the situation of local power bodies in the Soviet Union rather than in the European states' (Gelman 2007). We cannot fully agree with such sharp criticism, although it is obvious that with recentralization of the governance system in the country, the development of local self-government has slowed down. In an interview in 2005 one of the municipal leaders in the Murmansk Region said: 'Most important are the legal aspects – the federal level does not solve many legal problems and at the same time it does not give much room for the municipal level to solve these issues itself'. The most acute practical problem of local self-governments today is inadequate financing of the legislatively mandated powers at the municipal level and an increasing financial dependency of the municipal bodies upon the various forms of financial aid from the regional and federal levels (Didyk, Riabova and Ivanova 2007).

As to the relations between the regional and local levels of power, we would like to note the following. After 2000, important steps were taken in the regions by regional administrations to strengthen the position and development of municipalities. For example, in the Murmansk Region the most important initiatives of regional power bodies have been the adoption of regional laws on budget relations (relations between municipal and regional budgets) and the work on trilateral agreements (between the regional government, big industrial enterprises and municipal administrations) on social partnerships. However, in many regions (especially those relying on resource-based industries) regional power bodies do not prioritize the development of local self-governments. Our research shows that in the Murmansk Region the attention paid by the regional government to local self-government bodies is clearly less than that paid to big industrial enterprises or development problems, such as the development of the mining industry, fisheries and transport or the implementation of large projects such as oil and gas transportation. (Didyk, Riabova and Ivanova 2007).

Most of the interviewees were practically unanimous in their negative estimations of the development of local self-government in Russia and of its role in the general governance process. We have a more balanced opinion on this matter and consider that these overly pessimistic assessments can largely be explained as the application of high Western standards to what is a very young Russian local democracy, which leads to excessive expectations regarding local self-government in the Russian Federation. In our opinion, in 1990–2000 involvement of the local and municipal levels in the governance process has been noticeably increasing in Russia and this development has clearly provided important input into the rise of multilevel governance in the country. However, since 2000 there has been a shift in favour of stronger state influence on municipalities and more control over local self-governments by federal and regional power bodies. Such developments make us agree that today's system of local self-government in the country is not characterized by real municipal autonomy and is not fully based on real democratic practices.

One additional aspect of multilevel governance is the increasing involvement of (semi-)private actors in decision making. In this respect, it is obvious that the business and political elite in Russia today have become intertwined and, compared with the 1990s, the state bureaucracy has obtained more tools to exert influence on private business. A new economic elite has formed, mainly in export-oriented resource-based industries. Big industrial corporations are the principal actors representing business on the political scene. Representation of their interests often seems to be reduced to the interests of a narrow circle of owners and heads of companies, and the present situation is defined by some experts as oligarchic corporatism. The involvement of small- and medium-sized businesses in the governance processes is low and the influence of the related business associations ('Delovaya Rossiya' and 'Opora') on the decision-making process is much weaker than that of the Russian Union of Industrialists and Entrepreneurs ('Rossiskiy Soyuz Promyshlennikov i Predprinimatelei'), which represents big industrial corporations (Pshizova 2005).

An important sign of the rise of multilevel governance is the increasing significance of networks and the involvement of public authorities, NGOs, social movements and expert organizations. In this respect, Russia has definitely taken a step forward in recent decades. Development of civil society was quite intensive in the 1990s and 2000s, with close to zero participation by the state in this process. After 2000, the task of involving civil society institutions in governance processes at all of the levels was set at the national level, with Putin declaring in 2001 the goal of creating a maximally favourable environment for development of civil society in the Federation. One should remember that the bureaucratic machinery in Russia is a powerful force united by mutual administrative and corporative responsibilities that tries to keep a monopoly on information and managerial competence. 'Today our state is very reluctant to give up the habitual practice of straightforward supremacy; it is a reckless administration which is afraid of public and open dialogue with society. But the fact that it has started moving slowly towards recognition of the public importance of civil society, towards social partnership is obvious' (Nikovskaya 2004). The following facts are evidence of this trend: implementing national civil forums from 2001, the establishment in 2004 at the federal level of a presidential council to assist in the development of civil society institutions and human rights (transformed from the Commission on Human Rights established in 1993), the introduction of the Public Chamber of the Russian Federation and, in 2006, the creation of public chambers in the regions of the Russian Federation.

Today the state tries to learn to conduct dialogues with society. The circle of participants in decision making has broadened (largely at the expense of the involvement of expert organizations) and attempts have been seen to make connections with the real interests of various groups. Many scholars note that at the federal level new actors are involved in decision making in a 'consultancy regime' (Nikovskaya and Yakimets 2004). Many regions copy the same style. Opening new possibilities for civil society also creates new challenges. First, new demands are presented to its representatives. To function successfully in the 'consultancy regime', civil society has to develop strong competence in issues of state policy and to protect its public-political autonomy. Second, the consultancy regime increases the danger of strengthening bureaucratic corporatism. The traditional bureaucrats are not prepared for a regular dialogue with civic partners. The participation of external forces is still perceived as an encroachment on the sovereign territory of executive power and is often carried out formally.

Very recently, in the opinion of many experts, under conditions of the bureaucratic stabilization with control from administrative structures, the sociopolitical conditions for the formation of civil society in the Russian Federation and its role in public policy have narrowed. The main arguments supporting this view are that the sphere of public discourse of various public-political forces has been curtailed and that the system for forming political parties has become detached from the roots of civil society and implements only the function of an 'exhaust valve' that is fraught by stagnation of the political system. We do not share the most extreme pessimistic views on civic development in Russia. One should remember that it started to develop only very recently and is heavily burdened with the legacy of a totalitarian past. Here, too, a 'syndrome of over-expectation' can be identified. However, we agree that the 'civil vacuum' in Russia is not significantly decreasing. The lack of organized collective actions is evident. The significance of networks and the involvement of public authorities, non-governmental organizations, social movements and interest organizations that greatly increased in the 1990s has recently declined, although the involvement of expert organizations has increased. Today's degree of involvement of civil society in governance processes in the Russian Federation at all levels can be seen as unsatisfactory and only weakly supporting development towards multilevel government.

Transformation of Fishery Governance in Russia: The Case of the Murmansk Region

For most of the Soviet era, fishery management[1] was organized on three sector-based levels with strong vertical subordination. At the national level, the Soviet Ministry of Fisheries was the main regulation body. At the regional level, regional fishing-industrial associations, established in five fishery basins,[2] were the main sectoral bodies working under the ministry.

The Murmansk Region belongs to the Northern Basin, which covers the Murmansk and Arkhangel'sk Regions, the Republic of Karelia and the Nenets Autonomous Area. Fishing in this basin usually takes place on the Barents Sea. The Northern Basin is the second most important of the five basins, only the Far Eastern Basin being larger, and it has traditionally been a mainstay in the economy of Northwest Russia. The regional fishery association for the basin was *Sevryba*[3] with headquarters in Murmansk. It was a powerful organization that held assets, managed the fleet, organized the processing of fish and was responsible for the distribution of quotas among fishing organizations (Honneland, Ivanova and Nilssen 2003). Other regional bodies implementing the national fishery policies for more than 70 years were each basin's Department of Protection, Reproduction of Fish Stocks and Regulation of Fisheries (*rybvody*), which reported to the corresponding department of the Ministry of Fisheries in Moscow (*Glavrybvod*). The Northern Basin had three regional branches of Glavrybvod, one of which was *Murmanrybvod*:[4] Their functions were to licence fishing vessels, control quota

---

1 Here we use term 'management' to stress the strong hierarchical sector-based character of the governance process.

2 The five are the Far Eastern, the Northern, the Western, the Caspian Sea, and the Azov and Black Sea basins.

3 This is blend of the Russian words for 'north' and 'fish'.

4 *Murmanrybvod* in the Murmansk Region, Sevrybvod in the Archangel'sk Region and the Nenets Autonomous Area, and *Karelrybvod* in the Karelia Republic.

use and administer the closing and opening of fishing grounds. The lowest level of fishery governance was represented by fishing organizations (including fishery kolkhozy), which owned the ships. Thus, in that period fishery management was sector-based with strong regional sectoral actors that were vertically subordinate to corresponding structures at the national level.

In 1985, the established system entered a period of transformation. Gorbachev's ideas about the use of natural resources were based on the replacement of Soviet production goals by profit-and-loss paradigms. By 1988, the Ministry of Fisheries had ordered the regional fisheries associations to disband. The intent was to allow fishery interests in the regions to take over day-to-day economic management of the fleet (Pautske 1997). In 1989 every planned management structure of fisheries was changed. Under the new system, fishing fleets and land-based businesses could come or go freely from regional associations such as Sevryba.

Transformation of Fishery Governance and Changes in Federal-Regional Relations

In the spontaneous decentralization that the Russian Federation underwent in 1990–1993, the fishery sector experienced a period of collapse and dissolution of the former vertical sector-based Soviet system of management. After the breakdown of the USSR in 1991, an enormous change in the management of natural resources occurred. The remnants of the former system of fishery management were destroyed. The responsible ministry was abolished and the State Committee for Fisheries was established as its downgraded successor.[5] The federal bodies that have 'survived' in the basins are the rybvody and state administrations of the fishery ports (Honneland, Ivanova and Nilssen 2003; Vasiliev 2004). Regional associations such as Sevryba collapsed. A great number of small private fishery firms emerged on the ruins of huge state enterprises. In January 1992, fish prices were freed up. In December 1992, President Yeltsin signed a decree allowing fishers to keep up to 90 per cent of the hard currency earned from the export of fish products. Nonetheless, the rouble had been devalued, money was in short supply and the federal level was no longer able to provide sufficient fuel or money for fleet repair. The total annual catches in the Russian Federation rapidly fell off, declining from a peak of 10.3 million metric tons in 1988 to 4.3 tons in 1993 (Pautzke 1997). In the interviews, the fishery experts pointed out several main factors for such a drop: a decrease in the volumes of Total Available Catches (TAC), narrowed fishing areas due to the Russian fleet leaving remote oceanic areas and turning to fishing mainly in the Exclusive Economic Zone (EEZ) of the Russian Federation, high harbour payments for arrival of Russian vessels to homeports and inadequate customs policies (unreasonable customs duties for Russian vessels upon arrival

---

5   Ministries and state committees are different types of administrative bodies at the federal level. The ministries are placed higher in the political hierarchy. The state committees are not subordinate to any ministry.

at homeports). Together with the decline in catches, there began an enormous increase in deliveries abroad of Russian-caught fish .

The period **1993–2000**, identified for the country as an 'asymmetric contractual federation', could be called 'the period of centre-region battles for power' where fisheries were concerned. There were two main struggling actors. First, the State Committee for Fisheries was fighting to regain its former high status of a powerful body. Second, the administrations of the regions battled to secure their roles in fishery governance.

At the federal level, the State Committee for Fisheries fought to regain its former status as a powerful body on the ministerial level and to fend off interference from other federal bodies (such as Ministry of Economic Development and Trade, the Ministry of Natural Resources, the Ministry of Agriculture and various federal agencies) and from new regional actors. Before 2000, there were several attempts by other federal bodies to rescind the Committee's exclusive power to rule national fisheries. For example, in 1997 there was an attempt to deprive it of the status of a state committee and to incorporate it into the Ministry of Agriculture (the status was restored in 1998 due to general reorganization of the federal administration). Because of the upheaval in 1993–2000, the Committee worked out only one fundamental document regulating fisheries under the new market conditions, that is, the programme for the development of fisheries in the Russian Federation for the period up to the year of 2000, called 'Fish'. It was approved in 1995, but never fully implemented due to a severe lack of financing.

A new actor entered fishery governance at the regional level in 1991–1992 – the regional administrations. Thereafter, a real battle to strengthen the role of the regions in fishery governance began. In 1993–1994 fisheries departments were established under regional administrations in Murmansk, Arkhangel'sk and Karelia. In the USSR, regional authorities traditionally played a rather symbolic role in the governance of fisheries. During the 1990s their role increased significantly. This was an outcome, on the one hand, of Yeltsin's policy of delegating 'as much power to the regions as they can bear' in the federation treaties on the division of power between the federal and regional governments, in which regions formally acquired all jurisdiction over natural resources. On the other hand, it was a result of strong efforts by the regional authorities that have been often accompanied by open conflicts with the federal level. Sevryba in 1992 was re-established as a stock company consisting of twenty-three private firms. Thus, in the Murmansk Region two regional players – the regional administration and successor of the powerful regional fishery organization – started to demand recognition of their rights to influence fisheries governance.

The centre-region conflict was most intense in 1992–1996. In the Northern Basin, the regional administrations claimed their right to take part in fisheries governance in the country's EEZ, which had always been a federal responsibility. They demanded a say in quota distribution, then dominated by Sevryba through its participation in the Technical Scientific Catch Council. After resisting for a year, in 1993 the State Fisheries Committee handed over responsibility for the total Russian share of the Barents Sea quota to the administration of the Murmansk Region. Yet, this decision remained only partially implemented. Since 1994, the administration of the Murmansk Region has only distributed limited quota shares to the coastal fisheries (Eikeland, Riabova and Ivanova 2005). In 1997, the system was fully revised, with quota owners allowed to openly trade their quotas. In 1997–1998, Sevryba lost most of its regulatory responsibilities, which were divided among the regional fishery councils (responsible for intraregional quota allocation), largely controlled by the regional administrations, and Murmanrybvod (responsible for technical regulation). Control over the interregional quota allocation process was transferred from Sevryba to the State Committee for Fisheries (Honneland, Ivanova and Nilssen 2003; Honneleand 2005). Today all fishery experts admit that by 2000 the Murmansk Region, as well as other fishery regions, had 'lost the battle'.

The period **from 2000 to the present time**, 'the new centralism' for the country as a whole, can be called 'the return of fishery governance to the federal centre' for the fisheries. In 2000–2001 the Ministry of Economic Development and Trade introduced the auctioning of fishing quotas despite severe opposition from the State Committee for Fisheries and fishers (Honneland, Ivanova and Nilssen 2003). This led to a concentration of administrative, economic, political and other resources in the hands of the federal centre. At the end of 2003, the plan for the development of the fisheries of the Russian Federation for the period up to the year 2020 was adopted. It set the following goals: the creation of conditions for the increasing effectiveness of the export of fish products, optimization of fishery governance, achieving stable functioning of the fishery sector and enabling it to secure food self-sufficiency for the country, and fostering development of coastal fishery-based regions. The plan emphasized the return of Russian fisheries to international and open zones of the world oceans, cancelled fishing quota auctions, forbade distribution of quotas to foreign companies and allowed small fishery businesses to receive quotas for fishing in Russian coastal zones.

In 2004 a new, more centre-oriented phase in the restructuring of fishery governance began. In terms of number of reforms, it was an extraordinary year. In March 2004, President Putin introduced extensive changes in the federal administrative structure. The most important was the introduction of three categories of federal executive bodies: the federal ministries, the federal services, and the federal agencies.[6] The State Committee for Fisheries was abolished. Policy-making functions were transferred for the second time to the Ministry of Agriculture. Implementation of the state's fishery policies was handled by the newly established

---

6   The federal ministries define state policy and perform normative and legal regulation in their sphere of responsibility. The federal services carry out control and monitoring functions in their fields, or engage in protection of state borders or security more widely. The primary function of federal agencies is implementation of state policy and provision of services to the population. The last two are subordinate to a corresponding ministry (Honneland 2005).

Federal Fisheries Agency (FFA), which reported to the Ministry of Agriculture. Monitoring and enforcement (formerly functions of the old Glavrybvod) became the task of the federal service of veterinary control (Rosselhoznadzor under the Agriculture Ministry). The functions of the Glavrybvod were reduced and it became subordinate to the FFA.

Thus, three bodies of state executive power became responsible for fisheries governance: the Ministry of Agriculture, the FFA and Rosselhoznadzor (and their regional branches). In 2005, the Council on the Development of Fisheries under the Ministry of Agriculture was organized to coordinate the activities of these bodies. The main tasks of the Council were to assess related federal programmes and to make suggestions for developing Russian fisheries and their legislative basis.

At the regional level, the changes introduced in 2004 provoked numerous conflicts. One example is the conflict between Murmanrybvod and regional branch of Rosselhoznadzor. In 2005, due to a lack of or unclear orders on the federal level, these two bodies sharply disagreed in regard to licensing and controlling the fishing on the rivers of the Murmansk Region. For half a year, both had been carrying out the same functions, with Rosselhoznadzor forcefully demanding an exclusive right to licensing and controlling. In the interviews conducted with people working at the local offices of these bodies, as well as with fishers, a recurring theme has been complaints about 'a total mess as a result of the administrative changes'. Neither representatives of Murmanrybvod nor ordinary fishers appreciated the changes, showing little understanding of the logic behind what one of the experts called 'the destruction of old good system and ruining strong, competent and respected Murmanrybvod'. It was clearly stated in the interviews that the federal administrative reform in fishery governance had produced, at least at its initial stage, 'conflicts and the loss of experts and skills in the fisheries' regulatory bodies' at all levels.

The year 2004 was remarkable also in respect to the development of fisheries legislation. Before 2004 fishery activities were regulated by the Constitution of the Russian Federation, several agreements between federal and regional administrations and several federal laws on internal waters, the continental shelf and so on. However, the major legislative problem was the lack of a federal law on fisheries, which the State Duma had failed to adopt for many years. In 2004, the federal law titled 'On Fisheries and Conservation of Aquatic Biological Resources' was finally adopted after eight years of heated debate.

The new law became one more issue for a centre-region clash. It indicated the continuation of federal control over fisheries, including the economically important power of quota allocation. It minimized the power of the regions, contrary to the possibilities for regional control spelled out in the Constitution (Honneland, 2005). Even if the Constitution of 1993 did not fulfil all the expectations of the regions, it still made room for their participation in the governance of natural resources. According to article 72, 'possession, utilization and management of land, subsurface, water and other natural resources' fall under the joint jurisdiction of the Russian Federation and its regions (Constitution of the Russian Federation).

However, the new reforms returned the powers of the regions to the centre. In accordance with article 10 of the new law, fish and other aquatic bioresources are federal property[7] (Federal'nyi zakon No.166 2004). According to the law, the foundation for stabilizing Russian fisheries is the allocation of long-term fishing rights mainly based on the allotment of quota shares for a five-year period.[8] The law noticeably limited the rights of the regions, restricting their role mainly to giving advice about the quotas of coastal fisheries. Most of the interviewees pointed out that coastal fisheries became 'the only and the last hope for the regions to get control over fisheries'.

In 2005 the administrations of the fishery regions of the Russian Federation and the fishery associations started a new campaign aimed at regaining regional control over the sector. The target area became coastal fisheries, a sphere of vital importance for the fishing regions and an issue that was covered by only two lines in the new law (Federal'nyi zakon No.166 2004).[9] The Murmansk Region was very active in this process. In January 2005, the draft of the law 'On Coastal Fisheries' was delivered by the State Duma. The regions and fishery associations were invited by the federal centre to make their suggestions for amendments. In spring 2005 the administration of the Murmansk Region completed a list of suggestions, in fact, its own version of the law. The central point of the draft was a statement about power of the regions to distribute coastal quotas, which at the time was a privilege of the federal level. Other fishery regions demanded the same. The FFA formulated its position on such suggestions, which was categorically negative. As one of the FFA representatives said, 'the possibility to pass the power to distribute coastal quota to the regions should not be even discussed. Today's system allows regions to give advice about the quotas of coastal fisheries. The regions want more power in this area to protect their own narrow interests. This can result in an increase of illegal practices' (http://www.rbcdaily.ru).

By the time the collection of data for the contribution[10] was finalized, the regions of the Russian Federation had regulated their fisheries through the related departments in the regional administrations. In the Murmansk Region, this was the Department of Provisions, Fisheries and Agriculture, which has two subdivisions: Marine Fisheries and Investments and Development of Coastal Fisheries, Fish Processing and Aquaculture (http://www.gov-murman.ru/power/dep/food/). There were also many regional branches of federal structures involved in the governance of fisheries; it was not possible to obtain reliable information on how many there were in the Murmansk Region but it seems that they were quite numerous. In an

---

7   The exception is water biological resources of in-land water reservoirs, which can be federal, regional, municipal or private property.

8   Very recently it became 10-year period.

9   Article 20 of the law 'On Fisheries and Conservation of Aquatic Biological Resources' only states that coastal fishing is carried out by individual entrepreneurs and organizations registered in the regions (subjects) of the Russian Federation.

10   The data were collected by 2006.

interview at the end of 2005, the governor of the Murmansk Region, Y. Evdokimov, pointed out that 'state governance in the regions of the Russian Federation lacks flexibility as well as transparency and promptness. It is a result of the absence of an interconnected structure of the bodies of executive power. For example, in the Murmansk Region there are 20 executive bodies in the regional administration, 30 territorial bodies representing the executive power of the federal level and several departments of interregional state structures and federal state organizations. Many of them are even not known at the regional level' (http://www.hibiny.ru/news/).

The government of the Murmansk Region takes efforts to have a multiactor dialogue on maritime activities and establish cooperation between different actors on the regional level. In 2004, administration of the region organized the Regional Council on Maritime Activities, a collective advisory body consisting of representatives of regional administration, heads of self-governments of some coastal towns, representatives of the administration of the Murmansk fish port and non-commercial fishery associations, customs and border service officials, scientific experts, and so on. This was definitely a step towards the development of institutions of cooperation and coordination of a multilevel character.

By the time this chapter was written, the fishery governance functions in Russia were widely spread among the Ministry of Agriculture, the Federal Service of Veterinary and Phytosanitary Surveillance, the Federal Agency for Fisheries, the Ministry of Natural Resources, the Federal Service of Nature Management inspection, the Ministry of Economic Development, and more. Coordination was not well-established, and duplication of functions was often observed. The system did not allow for making practical decisions on time. Coordination of documents between various structures demanded much time. According to regional experts, all this hampered effective practical activities of the fisheries in the regions. Under such conditions, it is no wonder that since 2006 there has been discussion about re-establishing the Ministry of Fisheries. Both regional authorities and fishers have increasingly started to demand re-establishment of the ministry, their main argument being to 'have fishery governance in single hands'. Such a longing for the return of one sector-based governance body at the ministerial level points to a failure to reach well-coordinated cooperation between the various bodies of state power at the federal and regional levels.

Summarizing this part of the discussion and drawing on the broadest view of multilevel governance, we note that the increasingly active involvement of new regional actors – the administrations of the regions – in the fishery governance in 1993–2000 may at least point to possibilities of a turn towards multilevel governance in the fisheries of Northwest Russia. Although they both formally represent state power, federal and regional authorities were in fact strongly opposed. The most noticeable shift from the national to regional level in fishery governance took place in 1992–1996 with the declining importance of the nation-state level and, for the Murmansk Region, with the peak of regional participation in fishery governance in 1993. Since 1998 one has seen withdrawal from the region of most important functions (such as the responsibility for the total Russian share of the Barents sea quota) and devolution of less important functions to the regional level (intraregional quota allocation). General recentralization of governance in the Russian Federation resulted in restoration of a hierarchy in fishery governance dominated by the federal level. However, the new system differs from the Soviet model in that the functions are widely spread among the various state bodies, mainly the federal level. The new model with its multiactor format placed new, unfamiliar demands on its participants and a failure to reach well-coordinated cooperation between the various bodies of state power led to increasing demands from the regional level and fishers to re-establish the Ministry of Fisheries with all the power over fisheries concentrated in single hands.

The Changing Role of Local Self-Governments

What about the involvement of the local level in fishery governance? In the Murmansk Region, fisheries (the fishing fleet and fish-processing enterprises) are located in Murmansk (a city with more than three hundred thousand people) and in several fishing villages on the coasts of the Barents Sea and the White Sea.[11] Here we will focus on the involvement of local self-governments of fishing villages in fishery governance.

Since the beginning of the 1990s, the municipal level in Russia has been the object of changing political priorities of central actors. The priorities have shifted several times from favouring to neglecting. The hardest times were experienced by the self-governments of fishery villages of the Murmansk Region in 1993–1994. In 1993, the Municipalities Act adopted in 1991 was abolished after dramatic incidents in Moscow where Parliament and the Supreme Soviet clashed with the president. The period 1993–1994 before the adoption of a new Municipal Code in 1995 could be described as administrative uncertainty or even nullity in relation to municipal self-government. In the fishing villages of the Murmansk Region, it was a time of the most severe wave of socioeconomic crises. The newly established municipalities were left alone with responsibilities for local social and economic life and the formal freedom to act on their own. But this political decision was not supported economically, since the local level had the right to collect only some small charges and taxes. We found that during this period both sides perceived the relations between local administration and regional government as top-down and that this can be explained to a great extent by the legacy of the former centralized system. At the time the regional government had neither enough financial resources to sponsor the local level nor a conscious regional strategy for small resource-based communities. To understand whether and to what extent local self-governments of fishing villages

---

11   In the 1930s there were more than 80 fishing villages in the Murmansk Region. During the period 1930–2000, the number of fishery-based communities declined considerably (especially in the 1960s due to a policy of concentrating production). Today there are seven localities where fishery is officially recognized as a basic economic activity.

in Northwest Russia have been involved in fishery governance we will draw on the story of Teriberka, one of the fishing villages in the Murmansk Region.

Teriberka is a village of 1,400 inhabitants on the coast of the Kola Peninsula 120 km northeast of the city of Murmansk in the border zone of the Russian Federation (see the map at the beginning of the book). It is one of the few fishing villages left in this highly urbanized region. For centuries, it was a wealthy village and open to international contacts, especially in the times of Pomor trade before the Russian Revolution of 1917. During most of the Soviet period Teriberka experienced intensive development, having a population as high as 12,000 in the 1960s. In the last decades before *perestroika*, the population stabilized at 2,400 inhabitants. The economy was based on the fishing kolkhoz, which owned the fishing fleet, several fish-processing plants and a shipyard that was part of the powerful regional fishing complex 'Sevryba'.[12]

In the beginning of the 1990s, the village faced a deep socioeconomic crisis caused by the transformation of Russia to a market economy along with the introduction of strong regulation of fish resources. In 1993 the shipyard was closed down, and about half of the jobs in the fish processing plants of the kolkhoz were lost. The former 'pride of the coast' was about to collapse, as the local economy crashed. Living standards declined and a class of 'new poor' appeared, made up of well-educated people. The welfare infrastructure was cut back with a reduction of state transfers and a diminishing base for local financing.

Self-government of the village was established in the early 1990s and consisted of an elected mayor (head of local administration), secretary and bookkeeper employed in the municipal administration. Later, in 1996, a new representative body was introduced, the village council, consisting of nine elected people working for the council on a voluntary basis. The mayor played a very important role in this system, being in fact the only person administratively responsible for the socioeconomic situation in the village. During 1993–1994, the hardest years of crisis in the village, local self-government in Teriberka was mainly preoccupied with practical tasks related to the maintenance of social infrastructure in the village – securing the functioning of heating systems, schools, childcare and a hospital. The local economy became dominated by private fishery businesses, mainly managed from the offices in Murmansk. A number of newly established small private cooperatives operated nearby the village but most of them were organized by outsiders. Initial capital for establishing such cooperatives was needed and the local people lacked it. One of the villagers said in an interview: 'firms and

---

12   Teriberka has been studied in a number of MOST CCPP studies, including Riabova, L., 2001, 'Coping with Extinction, The Last Fishing Village on the Murman Coast', Aarsæther and Bærenholdt (eds) *The Reflexive North;* Skaptadóttir, Mørkøre and Riabova, 2001, Coping under Stress in Fisheries Communities, MOST Discussion Paper No. 53 and Skaptadottir, Mørkøre and Riabova, 2001, 'Overcoming Crisis: Coping Stratigies in Fishery Based Localities in Iceland, North-Western Russia and the Faroe Islands', in Bærenholdt and Aarsæther (eds): *Transforming the Local.*

cooperatives rob the coast, and we have nothing'. At that time fish poaching (both organized poaching taking criminal forms and individual cases as people's way of coping with poverty) became a serious problem and remains so today, having negative effects both socially and environmentally.

In 1993–1996, regional power over fisheries was comparatively strong. In 1994, the government of the Murmansk Region started to implement stronger policies towards fishery-based communities. The dominant strategy in Teriberka was to attract external (primarily foreign) capital to the traditional sector of its economy. These efforts resulted in establishing a joint Russian-Portuguese-Lithuanian fish processing enterprise and a cooperative project with the Norwegian municipality of Båtsfjord. These business initiatives were mainly a result of ties between outside entrepreneurs who were working on the basis of long-distance commuting or from offices in Murmansk and who either formerly worked in Teriberka or knew about the place from business partners. The local and regional governments started to cooperate and jointly provided political and administrative support for measures to attract external resources and to stimulate cross-border cooperation. Since the regional government at that time was responsible for distributing quota shares to the coastal fisheries, local self-government had a strong influence on coastal quota distribution. In addition, local self-government at that time could influence the development of fisheries through a quite flexible local taxation policy that included establishing extra taxes or tax preferences locally for certain kinds of activities (Riabova 2001).

New initiatives in fisheries have been important for keeping the local economy alive. In the early 2000s, the people of the village had a possibility to work in fish processing both at the factory that belonged to the kolkhoz and at the new enterprise. In addition, 40–50 workers from Teriberka (9 per cent of the workforce) were employed in fish processing in Båtsfjord, which was seen as a fortunate way to cope with poverty. However, local economic life has not recovered yet and the village is still struggling to come to terms with the economic transformation and the crisis in local fisheries. About half of the workforce is still without a job.

Recentralization of fishery management, which started in 2000 and strengthened in 2005 after the adoption in late 2004 of the federal law 'On Fisheries…', considerably reduced possibilities for the involvement of local self-government in fishery governance. The new law transferred the responsibilities for coastal quota distribution to the federal level, which automatically meant depriving local self-governments of influence on this process. The law includes such principles as participation of the local population in decision making regarding fisheries and possibilities for the local self-government to issue its own legislative documents regarding fisheries. However, the law does not suggest concrete mechanisms for implementing such principles and in practice development of local legislation is strongly restricted by federal laws. Moreover, after the late 1990s it became increasingly difficult to get permission from the military authorities for foreign business activities in the border villages, reducing foreign business activity in Teriberka. In addition, with the adoption in 1997 of the new Budget Code of the

Russian Federation, local self-governments lost most of its rights to formulate local taxation policy.

Since 2004, regional and municipal governments in the region have focused their efforts on providing legislative and economic support for coastal fisheries. In 2004, the regional programme 'Development of coastal fisheries, aquaculture and fisheries in inland water reservoirs for 2004–2008' was adopted. The programme focused mainly on the expansion of cooperation at the regional level with business structures in order to develop the sector, while the municipal level seems to have a secondary role in these plans. Surprisingly, heads of small fishery-based communities were not invited to participate in the Regional Council on Maritime Activities.

Recent oil and gas transportation developments in the region bring another perspective on life in Teriberka. There are prospects that a gas processing plant (LNG terminal) for the Stockman field will be constructed in the village by the Russian energy giant Gazprom. In addition, three platforms will be placed 560 km off the coast of the Kola Peninsula. An undersea gas-extracting complex will be placed at a depth of 340 m, from which three pipelines will be connected to Teriberka (http://barentsobserver.com/shtokman-project-presented-to-the-people-of-teriberka.4473236-16149.html). According to plans, the plant is to be finished by 2014. At the public hearings in the village, the Gazprom representatives admitted that despite adhering to international standards the project will inevitably affect the environment and fishing in the area. However, most likely the local self-government and villagers will have to consent to the project plans. Gazprom has prepared a package of offers, which most of the local population is unlikely to resist even though some of the villagers desperately want to preserve the place as a fishing village and to continue a fishery-based way of living. It has been promised that new houses will be built, jobs created and service industries developed. It is highly likely that the most famous fishing village in the region, the former 'pride of the coast', will in a few years be transformed into an industrial town and a key place for Russian oil workers.

The Teriberka story shows that for the last decade the functions of local self-government in Russian fishery management have been narrowing. In the mid-1990s local self-governments had broader opportunities to take part in fishery governance, the most important of which was involvement in coastal quota distribution and the possibility to form local taxation policy in order to stimulate business development in the communities. Local self-governments in Northwest Russia, like those in other maritime regions are left with very limited possibilities to influence fishery development and fight the crisis in local fisheries. Thus, coastal communities dependent upon fisheries must contend with a future that appears vague at best.

The Role of Private Industry

Another aspect of multilevel governance is the increasing involvement of (semi-) private actors in decision making. The number of private firms in the fishing industry exploded in the 1990s. At present over 90 per cent of the fishery companies in the

Murmansk Region are private; there is no direct management of them by the state. The activities of the private actors are regulated by the state indirectly, on the basis of the civil, tax, customs and other legislation. In 2002, there were 210 fishing and fish-processing companies, of which 102 were engaged in marine fishing, 65 in coastal fishing, 26 in both fishing and fish processing and the rest, 43, in fish processing (Evdokimov 2002). In addition, there were fishery kolkhozy (collective farms) in the region; these numbered 41 in 2005 (http://www.murman.ru/themes/economy-25112005(3).sh) The main problems that the new private fishery businesses have had to cope with include discontinuation of state subsidies, fishing vessels' expanded capacities, rising operating costs, loans and unreasonable and confiscatory tax and custom legislation in the home country. To operate effectively and be economically viable, they have had to use relatively large volumes of quotas or fish illegally if they did not have the resources to buy enough quota shares legally.

To defend their interests, representatives of the fishery businesses in the Murmansk Region have united in organizations and unions. The leading organizations of the region are the 'Union of Fishery Organizations' (a Russia-wide organization), the consortium 'Murmansk Trawl Fleet', the non-commercial organization 'Union of Fish Industrialists of the North', the non-commercial union 'Association of Coastal Fish Industrialists and Farms of Murman'. Fishery kolkhozy are united in the 'Union of Fishing Kolkhozy of the Murmansk Region' (part of a Russia-wide organization). The main functions of these organizations are to support the production activities of their members and to protect the rights and interests of their members in negotiations with bodies having power at the federal and regional levels and local self-government. All of these organizations are very active on both the regional and national levels.

One of the most active and influential organizations is the 'Union of Fish Industrialists of the North ', led by Gennady Stepahno. This non-commercial union was established in 1992 and today it has become one of the biggest associations of small and medium-sized fishery businesses in the Russian North. In 2003, the organization comprised 96 fishery enterprises and firms (almost half of the total number in the Murmansk Region), of which 73 are involved in fishing. The union's members own a total of 116 fishing vessels and 13 transport ships. About 7,200 persons were employed in the union's enterprises and firms, that is, half of the total number of persons employed in the region's fisheries (Stepahno 2003).

The union is well known for its energetic efforts in solving the practical problems of the northern fisheries and for its independent and persistent policies at both the national and regional levels to improve fishery governance. At the national level, it regularly takes part in discussions at all-Russia forums dealing with the transformation of fishery governance in the country. It also approaches federal power bodies (government of the Russian Federation, federal fishery-regulating bodies and the State Duma) with claims and proposals for improving the performance and prospects of the fishing industry as well as the general improvement of fishery governance. On the regional level, the union conducts activities aimed at establishing regular cross-sectoral or multilevel dialogue on regional fisheries. In the interview G. Stepahno said: 'Each month we conduct so-called "harvest hour" meetings with representatives of different organizations related to fisheries, such as the administration of the Murmansk fish port, scientific experts, customs and border service officials'.

However, although the associations of private fishery businesses in the Murmansk Region are strong and active, and have achieved certain practical results in their efforts, their role in fishery governance is quite modest. As one of the most experienced fishery experts in the region, V. Zilanov, expressed it, 'unfortunately, the role of non-commercial fishery associations is extremely modest, both in day-to-day decision making and in forming long-term fishery policy in Russia. Their opinions are not taken into consideration at high levels. Most decisions are made at the federal level in an administrative manner and they very often do not facilitate sustainable development of Russian fisheries' (Zilanov 2002). The union is the only non-commercial association in the region that is represented on the Regional Council on Maritime Activities (a collective advisory body organized by the administration of the Murmansk Region).

The Role of Environmental NGOs

There are many non-governmental environmental organizations in the Murmansk region, for instance, the Kola Coordinative Environmental Centre 'GAIA', the Kola Wild Nature Protection Centre, and the Barents Sea project office of WWF Russia and 'Nature and youth'. None of them, except the WWF, has participated in the projects dealing with marine fisheries, which was involved in the project on sustainable fisheries after establishing its office in Murmansk in 2004. Unfortunately, we do not have information on how this project has been realized. In the mid-2000s, GAIA had positive experiences of cooperation with the regional administration in cleaning small salmon rivers on the Kola Peninsula. After GAIA presented the project to the regional administration, a decision was taken to partly finance the project on the regional level.

According to the representatives of ecological NGOs' opinions, the regional administration mainly expects these organizations to deal with ecological information and education, not to take part in decision making. For example, in the Regional Council on Maritime Activities there are no representatives of ecological NGOs. Thus, we assess the participation of ecological NGOs in fishery governance in the Murmansk Region as being extremely low. We do not have enough information to estimate the participation of other public non-governmental organization, but we suppose that the situation is quite similar.

The Participation of Expert Organizations

The involvement of expert organizations in fishery governance in the Murmansk region mainly consists of the participation of two scientific organizations – the Polar Research Institute of Marine Fisheries and Oceanography (PINRO, director

Boris Prischepa) and the Murmansk Marine Biological Institute (MMBI, director Gennady Matishov). PINRO has the status of a federal unitary enterprise, while MMBI belongs to the Kola Science Centre of the Russian Academy of Sciences (KSC RAS). The main task of PINRO is to collect data and make yearly and long-term prognoses for harvesting aquatic biological resources in the Barents Sea and the North Atlantic. PINRO sends recommendations to the corresponding institute in Moscow and the proposals are then used for quota decisions on the federal level. On the regional level, the recommendations of PINRO serve as a basis for day-to-day and long-term planning of the development of fisheries in the Northern Basin. MMBI conducts basic scientific research and its recommendations are also taken into consideration, although on a more flexible basis.

Although fishing stock assessment is an inexact science in any country, in Russia fishery expert organizations often become targets for extremely critical 'shooting' from all sides in a process as highly politicized as decisions on stock shares. Representatives of the fishery businesses accuse scientific experts of taking an overly cautious approach and imposing too many restrictions on harvesting possibilities. On the contrary, ecological organizations and journalists tend to make scientific experts at least partly responsible for overfishing. Both parties unanimously consider that scientific recommendations are highly dependent on funding sources and future prospects for financing and are thus inadequate or doubtful. As one of the business persons said in the interview, 'we don't understand what the basis for quota decisions is. Is it science only, or is it a bit of science and a lot of commerce and politics?'

One way or another, the involvement of regional fishery science organizations and experts in fishery governance at both the national and regional levels is quite strong, although not equally represented in respect to field of science. For example, there is also the Institute for Economic Studies KSC RAS, which does research on economic and social aspects of fishery governance, regionally, nationally and internationally. Experts from the Institute are invited from time to time to take part in discussions on fishery development at the regional level, but they are not involved in regional fishery governance on a regular basis. Both PINRO and MMBI have representatives on the Regional Council on Maritime Activities, but there are no representatives of IES on the Council. Thus, participation of experts specialized in biological aspects of fishery science is favoured.

Conclusion

The aim of this study was to discuss the development of fishery governance in Norwest Russia from the multilevel governance perspective. The concept refers to non-state governance or joint public-private governance. However, the development of fishery governance in Russia, like that of any other sphere, is based to a considerable extent on relations between the federal centre and the regions. Being interested in studying fishery governance through the concept of multilevel governance, we have suggested that in extremely centralized countries like Russia the process whereby the centre loses its power to the subnational (regional) level is an important step towards the rise of non-state governance. It can be seen as a process constituting the development of multilevel government and thus centre-region relations may be included in the analysis.

Drawing on this view, we found that there were two major periods in the recent development of fishery governance in Russia: 1993–2000 and 2000 to the present. The years 1993–2000 were a time of centre-regions battles for power over fisheries. The period since 2000 is characterized by the returning of fishery governance to the federal centre. Since the beginning of the 1990s, federal and regional authorities, although both formally representing state power, were in fact strongly opposed to one another in the course of the general decentralization process. The increased involvement in fishery governance of the administrations of the maritime regions in Northwest Russia, especially that of the Murmansk Region, culminated in their acquiring substantial powers over fisheries in the mid-1990s; among other things, the government of the Murmansk Region gained a prominent role in distributing fishing quotas for the Barents Sea and coastal waters. This development can be understand as a step that opened up opportunities for the rise of multilevel governance in the fisheries of Northwest Russia. Non-state actors, such as local self-governments, private business and non-governmental public organizations had relatively extensive opportunities to take part in fishery governance in the course of decentralization. Local self-governments of fishery-based communities had a noticeable influence on coastal quota distribution, as the regional governments were responsible for coastal quotas and could influence the development of fisheries through quite flexible local taxation policies, including locally established taxes or tax preferences for fishery activities. Private actors, although experiencing a difficult 'pioneer' period of development, managed to establish several non-commercial associations, for example, the 'Union of Fish Industrialists of the North' in the Murmansk Region, and became very active in discussions of the development of national and regional fisheries. Ecological NGOs in the Murmansk Region have not been active in regard to fisheries, but expert organizations – PINRO (Polar Research Institute of Marine Fisheries and Oceanography) being the most prominent – have played a very important role regionally and nationally in decisions on fishing quotas. All these trends taken together, in our opinion, constitute a shift towards – or at least an opening for – multilevel governance in the governance framework of fishery in the Russian European North.

However, the hasty, 'right to the bottom' decentralization that occurred did not have the proper institutional, economic or social preparation. We believe that decentralization of governance can have positive economic and social effects only under two fundamental conditions. First, the former system of centralized control over activities of regional and local governments has to be replaced by a system of democratic control at the regional and local levels. Second, there has to be established a well-functioning regime of cooperation and coordination between various bodies of state power and other actors such as private business, public and expert organizations. Neither the first nor the second precondition in Russia was fulfilled. Democratic mechanisms of control over fisheries were formed too slowly. New non-state actors were hardly able to establish cooperation and a multiactor dialogue. Coordination between state and non-state actors was poor. The two main levels of power (federal and regional) in regard to such strategic resources as fish demonstrated mainly rival instead of cooperative behaviour. Equilibrium between levels of power based on rival behaviour usually leads to reduced effectiveness of the sphere where it is implemented. This is what happened in the case of the fisheries in the Northwest. They underwent a systemic crisis that has had extremely negative socioeconomic consequences undermining the economic health of private fishery businesses, the well-being of fishery-based communities and the economic wealth of the Russian Federation as a whole.

According to federal authorities, the new centralization was the reaction of the federal level to the low socioeconomic effectiveness of decentralization. The general recentralization of governance in the Russian Federation that has occurred since 2000 has led to a restoration of a hierarchy in fishery governance dominated by the federal level. However, unlike in the Soviet model, the functions are spread widely among various state bodies, mainly those on the federal level. The new model with its multiactor format put demands on its participants in terms of cooperation and coordination. However, lack of cooperation and coordination can also be seen observed in the new model, although today it is primarily one level – the federal – that takes the most important decisions on fisheries. The roles of other actors – regional governance and non-state – has noticeably diminished. In the Murmansk Region, like in other maritime regions of the Russian Federation, the role of regional governments and local self-governments in fishery governance since 2000 has been reduced considerably. Today it is only coastal fishery that regional administrations can influence to a certain extent. The functions and opportunities of local self-government in regard to fishery management have narrowed. Associations of private businesses in fisheries in the Murmansk Region are strong and active, and have achieved certain practical results in their efforts, yet their role in fishery governance is quite modest since higher levels of power demonstrate minimal responsiveness to the associations' proposals. We assess the participation of ecological NGOs in fishery governance in the Murmansk Region as extremely low. Involvement of fishery science organizations and experts in fishery governance both at the national and regional levels is quite strong, but they are not equally represented in respect to field of science: participation of experts specialized in biological aspects of fishery is preferred, while experts on socioeconomic issues are less involved in the consultancy process.

Regionally, some efforts have been observed to create a multiactor dialogue on maritime activities, to establish cooperation and to improve coordination between different fishery actors. One such effort is the initiative of the Union of Fish Industrialists of the North known as the 'harvest hour', which consists of monthly meetings with representatives of different organizations related to fisheries such as the administration of the Murmansk fish port, scientific experts, and customs and border service officials. Another initiative, put forward by the government of the Murmansk Region, is the Regional Council on Maritime Activities – a collective advisory body consisting of representatives of the regional administration, heads of self-governments of some coastal towns, representatives of the administration of the Murmansk fish port, non-commercial fishery associations, customs and border service officials and scientific experts. This is definitely a step towards the development of institutions of cooperation and coordination of a multilevel character. However, it is only partly a multilevel body: small fishery-based communities and ecological organizations have not been invited to send representatives to the Council.

Thus, the most recent developments show that fishery governance in Russia today has turned in a direction quite opposite to that required for multilevel governance. Efforts taken regionally, like those in the Murmansk Region of Northwest Russia, towards establishing multilevel dialogue and cooperation in regional fisheries, have, as one of the experts said, 'the impact of a drop in the ocean'. However, we choose to believe that with time drops cut stones.

References

Bache, I. and Flinders, M. (2004), 'Themes and Issues in Multi-level Governance', in Bache and Flinders (eds).

Bache, I. and Flinders, M. (eds) (2004), *Multilevel Governance* (Oxford: Oxford University Press).

Bærenholdt, J.O. and Aarsæther, N. (eds) (2001), *The Reflexive North* (Copenhagen: Nordic Council of Ministers).

Didyk, V., Riabova, L. and Ivanova, L. (2007), 'Regional-municipal relationship: the case of Kirovsk municipality'. Paper for the *Citizenship and Democracy: Networking Local Governance workshop*, Tromso.

Eikeland, S., Riabova, L. and Ivanova, L. (2005), 'Northwest Russian Fisheries after the Disintegration of the USSR: Market Structure and Spatial Impacts', *Polar Geography* 29:3, 224–36.

Evdokimov, Y. (2002), 'Kolskiy Krai: Samodostatochniy Rybopromishlenniy Region' (Kola Region is a self-providing fish producing region), Rybniye Resursi (*Fish Resources*) 1, 2–7.

Federal'nyi Zakon No.166 ot 20 Dekabrya 2004 g., 'O Rybolovstve i Sohranenii vodnyh Biologicheskih Resursov (v redaktsii ot 31.12.2005, No.199 FZ)' (Federal Law No. 166 FZ of December 20, 2004 'On Fishing and Preservation of Aquatic Biological Resources (reduction of 31.12.2005, No.199 FZ) <http://www.nature.ykt.ru/CD_zakon/rf%20zakon.htm>.

Gelman, V. 'Ot Mestnogo Samoupravleniya – k Verticali Vlasti', (From local self-government – to vertical line of power), *Polit.ru* [website], (updated 16 Apr. 2007) <http://www.polit.ru/research/2007/04/16/gelman.html>.

Honneland, G. (2005), 'Towards a Precautionary Fisheries Management in Russia?', *Ocean & Coastal Management* 48(7–8): 619–31.

Honneland, G., Ivanova, L. and Nilssen, F. (2003), 'Russia's Northern Fishery Basin: Trends in Regulation, Fleet, and Industry', *Polar Geography* 27:3, 225–39.

'Iz-za Vysokih Tsen na Toplivo Rybolovnyi Promysel Prinosit Bol'hsie Ubytki' (Fisheries suffer from losses due to high prices for fuel) *Murman.ru* [website], (updated 25 Nov. 2005) <http://www.murman.ru/themes/economy-25112005(3).sh>.

Konstitutsiya Rossiiskoi Federatsii (Constitution of Russian Federation), <http://www.constitution.ru/>.

Lapina, N. and Chirikova, A. (2004), 'Putinskie Reformy i Potentsial Vliyaniya Regional'nyh Elit'(Putin's reforms and potential of influence of regional elites), <http://www.fesmos.ru/Pubikat/19_RegElit2004/regelit_4.html>.

Larsen, L.-H. Northern Maritime Corridor: Inter Trade Business Trip to Murmansk and Archangelsk 21–26 March 2004. Summary of presentations and notes. *Northernmaritimecorridor.no* [website], (updated 31 Mar. 2004) <www.northernmaritimecorridor.no/ir/file_public/download/Nm.0Lars-Henrik%20Larsen.pdf>.

Nechaev, V. (2008), 'Tsentralizatsiya Versus Detsentralizatsiya: Sovremennaya Rossiya na Puti k Effektivnomu Gosudarstvennomu Upraleniyu' (Centralization versus decentralization: contemporary Russia on the way to effective state governance) (30 May 2008) <http://elis.pstu.ac.ru/nechaev3.htm>.

Nikovskaya, L. (2004), 'Ne osvoiv kul'turu konflikta, rossiiskoe obschestvo ne stanet grazsdanskim' (If Russian society does not learn culture of conflict, it will not become a civic one), *Novopol.ru* [website], (updated 30 Aug. 2004) <http://www.novopol.ru/article309.html>.

Nikovskaya, L. and Yakimets, V. (2004), 'Publichnaya Politika v Usloviyah Rossiiskoi Transformatsii' (Public policy under the conditions of Russian transformation), *Novopol.ru* [website], (updated 21 Sept. 2004) <http://www.novopol.ru/article491.html>.

Pautzke, C. (1997), 'Russian Far East Fisheries Management', (30 Sept. 1997) <http://www.fakr.noaa.gov/npfmc/summary_reports/rfe-all.htm>.

Peters, B. (1996), *The Future of Governing: Four Emerging Models* (Lawrence, Kansas: University Press of Kansas).

Pierre, J. and Peters, B. (2000), *Governance, Politics and the State* (London: MacMillan).

'Printsipy Zakrepleniya Polnomochiy za Regionami Chetko ne Ustanovleny: Interview Gubernatora Murmanskoi oblasti' (Principles of fixing the rights of the regions are not clearly defined) *Hibiny.ru* [website], (updated 30 Nov. 2005) <http://www.hibiny.ru/news/news.php?id=1176>.

Pshizova, S. (2005), 'Bisnes kak Gruppa Interesov v Politicheskoi Systeme Sovremennoi Rossii' (Business as a system of interests in the political system of contemporary Russia), *Vlast'* 2, 26–9.

'Regiony Tyanut 'Rybnoe Odeyalo' na Sebya' (Regions pull 'fish blanket' to themselves), *Fiveocean.ru* [website], (updated 13 Sep. 2005) <http://www.rbcdaily.ru>.

Rhodes, R. (1997), *Understanding Governance. Policy Networks, Governance, Reflexivity and Accountability* (Maidenhead: Open University Press).

Riabova, L. (2001), 'Coping with Extinction: the Last Fishing Village on the Murman Coast', in Bærenholdt & Aarsæther (eds).

Stephano, G. 'Vmeste My – Sila!' (Together we are strong!) *Fishres.ru* [website], (updated 25 Mar. 2003) <http://www.fishres.ru/cgi-bin/news/print.cgi?num=03115>.

'Stockman project presented to the people of Teriberka', *Barentsobserver.com* [website], (updated 8 Apr. 2008) <http://barentsobserver.com/shtokman-project-presented-to-the-people-of teriberka.4473236-16149.html>.

Stoker, G. (1998), 'Governance as Theory: Five Propositions', *International Social Science Journal* 50:1, 17–28.

'Structura Departamenta Rybnoi Promyshlennosti Murmanskoi Oblasti' (Structure of Department of Fisheries of the Murmansk Region) *Gov-Murman.ru* [website],<http://www.gov-murman.ru/power/dep/food/>.

Turovskiy, R. 'Tsentralizatsiya Rossii: Tsykly Prostranstvennogo Razvitiya' (Centralization of Russia: circles of spatial development), *Rustrana* [website], (updated 10 Mar. 2006) <http://rustrana.ru/article.php?nid=19784&sq=19,22, 613,1960&crypt=#n>.

Vasiliev, A. (2004), *'Teoreticheskie osnovy povysheniya effektivnosti funktsionirovaniya rybnoi otrasli na Evropeiskom Severe Rossii' (Theoretical basis of functioning of fishery industry of the Russian European North)* (Apatity: Kola Science Center RAS).

Zilanov, V. 'Rybnaya Promyshlennost' Sosednei Norvegii: Izuchaem Opyt' (Fishery industry of neighboring Norway: learning experience) Npacific.ru [website], (updated 25 Dec. 2002) <http://www.npacific.ru/np/gazeta/2002/1/tv7_58_04>.

Chapter 6

The Struggle for the Ownership
of Pulp and Paper Mills

Anna-Maija Matilainen

Introduction

The present chapter deals with the pulp and paper industry in Northwest Russia and its changing patterns of ownership. In recent years, the industry has undergone heavy restructuring and consolidation, with some producers starting to form holding companies to strengthen their position on the markets. In addition, large foreign forest enterprises have discovered the opportunities in Russia's forest sector and many have made significant inroads into the country. The pulp and paper industry has also attracted the interest of Russian oligarchs and their more active role in the business has caused a great deal of confusion. Since the beginning of the 2000s, the industry has been shaken up by a wave of enterprise takeovers. Forest enterprises are becoming concentrated into larger units and more powerful enterprises are emerging in the sector. Large and relatively lucrative pulp and paper mills have been the main targets of takeover attempts.[1] Accordingly, takeover battles have had a great influence on the entire sector. Furthermore, lengthy and complicated takeover battles have generated harsh criticism of and lively debate on the possibilities to regulate enterprise takeovers.

In this contribution I look at the actors in the pulp and paper industry in Northwest Russia and some prominent ownership disputes over pulp and paper mills. Russian newspaper articles in both national and regional newspapers have provided me information on the ongoing debate on takeover battles. I examine the Russian courts as well, since the defects in the Russian court system and the laws regulating court proceedings, as well as rampant corruption, have offered fertile ground for ownership struggles. In addition, the interviews conducted during the project Governance of Renewable Natural Resources in Northwest Russia have provided useful background information on the pulp and paper business in Northwest Russia. My particular focus here is on the role of law in takeover battles. Laws, particularly the law on insolvency and the law on joint-stock companies, have been used rather creatively in takeover attempts in the pulp and paper industry. For example, there are several cases in which an enterprise

---

1   In this chapter I review ongoing ownership disputes affecting Russian pulp and paper mills until summer 2007.

takeover was the underlying motive for a bankruptcy petition. In particular, the 1998 law on insolvency was widely used as an enterprise takeover instrument until 2002, when a new law came into force. Companies seeking to seize control of other companies have also taken advantage of minority shareholders' rights in the Russian law on joint-stock companies.

US and European models have had considerable influence on the Russian law on insolvency and law on joint-stock companies. Borrowing legal rules from one legal system to another has been discussed widely in comparative law and the research dealing with legal transplants is extensive. The debate is connected with the interaction between law and society and how legal systems evolve in relation to one another. In a nutshell, the main question is whether laws can be easily transferred from one country to another. For instance, Alan Watson suggests that legal rules, norms and systems may be successfully borrowed even where the circumstances of the host, or recipient, are different from those of the donor. He contests the theories that regard law as a mirror of society and suggests that changes in legal systems are based on legal transplants (Watson 1993). Pierre Legrand, on the other hand, is more sceptical and stresses that the idiosyncrasies of diverse legal cultures make borrowing legal rules from another legal systems complex (Legrand 1996; Teubner 1998). There are different opinions on whether the term 'legal transplant' describes the phenomenon appropriately.[2] According to Gunther Teubner, the term 'legal irritant' expresses things better than 'legal transplant', which creates the false impression that the foreign rule remains unchanged when applied in new conditions. He suggests that some areas of law are rather strongly coupled to social processes and that the degree of success of a transplant depends on the degree of coupling (Teubner 1998).

Some Main Actors in the Pulp and Paper Industry in Northwest Russia

During the Soviet era, the state controlled the pulp and paper industry as well as other industrial sectors. The industrial policy of the Soviet Union was predominantly oriented towards development of large-scale production with the objective of self-sufficiency within the domestic market and total employment (Dudarev et al. 2002). The term 'forest industry complex' came into use after World War II to refer to production complexes comprising mechanical and chemical forest industry units. The basis of such complexes was a pulp and paper mill accompanied by sawmills and other wood-processing plants (Eronen 1999). The Russian pulp and paper industry faced totally new conditions when the Soviet Union collapsed. Forest enterprises were privatized and several new actors came into the Russian pulp and paper business. Moreover, enterprises could no longer

---

2   In order to keep my focus clear, I have not entered the terminological debate in detail. Differences notwithstanding, the debate is fundamentally about the same phenomenon – borrowing legal rules and models from other legal systems.

rely on the state to find markets and provide investment capital (Kortelainen and Kotilainen 2003). My purpose here is not to list exhaustively all the pulp and paper producers that operate in Northwest Russia but merely to give some examples. Four different groups of actors can be identified in the industry: 1) company managers and employees, 2) Russian holding companies, 3) Russian oligarchs, and 4) foreign enterprises (Kortelainen and Kotilainen 2003, see also chapter 8 of this volume).

Most of the large Russian pulp and paper mills were privatized in 1993 and 1994. The period was a difficult one for the mills and the adverse situation on the world market made them uninteresting to both foreign timber companies and Russian investors. In most cases, employees and managers became the owners of newly privatized combines (Butrin 2004b). Privatization programmes gave certain privileges to the company insiders. The OAO[3] Kondopoga pulp and paper mill in the Republic of Karelia and the OAO Solombala pulp and paper mill in the Arkhangel'sk Region are good examples of successful pulp and paper combines that have remained under the control of company managers and employees. The Kondopoga mill, located in the town of Kondopoga, is a significant newsprint producer. The privatization of the mill in 1994 was peaceful and 80 per cent of the company shares went to the labour collective (Butrin 2004b). The German pulp and paper company Conrad Jacobson Group is a major foreign shareholder of the company. The Republic of Karelia also owns company shares. The town of Kondopoga is heavily dependent on OAO Kondopoga, as became evident during interviews made in Kondopoga and Petrozavodsk in autumn 2004. (For more about the interaction between company and town, see Chapter 8.) The OAO Solombala mill is located just outside of the city of Arkhangel'sk. It is specialized in manufacturing and exporting high-quality unbleached sulphate pulp (Solombala pulp and paper mill). The combine has remained under control of its managers and employees since it was privatized and converted into an open joint-stock company in the early 1990s (Butrin 2004b). Recently, the mill has come under the control of a new holding company, Solombalales (Dimitrev 2007). Solombalales includes a sawmill and wood-processing enterprise, several logging companies and other enterprises that support the operations of the holding company. (Solombala pulp and paper mill) Both the Solombala and Kondopoga mills have managed to avoid ownership struggles so far. However, in summer 2004 market rumours suggested that the company Basic Element had offered to buy the Solombala mill (Antanta Capital 2006).

The privatization of Russian forest enterprises severed the traditional ties within the industry. Even combines that previously had been part of a larger complex were separated (Kortelainen et al. 2003; Dudarev et al. 2002). Recently, the ownership of Russian forest enterprises has become more and more concentrated and many enterprises have started to build holding companies. Usually these holding

---

3   Open joint-stock company. The abbreviation 'OAO' comes from the Russian term *otkrytoe aktsionernoe obshchestvo.*

companies comprise one or more pulp and paper mills, suppliers of raw materials, other wood-processing enterprises and enterprises which are responsible for the distribution of end products. Ilim Pulp Enterprise, the Arkhangel'sk pulp and paper mill and Continental Management can be mentioned here as examples of large holding companies operating in Northwest Russia. New holding companies are being established in the Russian forest sector as well. Investlesprom is one example of the newcomers in the pulp and paper industry. The holding company operates in Northwest Russia and comprises several pulp and paper mills, logging companies, sawmills and other wood-processing enterprises. For instance, the Segezha pulp and paper mill in the Republic of Karelia is an essential part of the holding company (Investlesprom).

Initially, Ilim Pulp Enterprise specialized in the distribution of the products of the Ust-Ilimsk and Kotlas combines. Soon it became clear that business stability and efficiency would depend on a consolidation of production processes, and the company adopted a policy of building a single vertically integrated company comprising, among other operations, logging, distribution and marketing (Ilim Pulp Enterprise). Ilim Pulp Enterprise controls some large pulp and paper combines, such as the Kotlas mill in Koryashma in the Arkhangel'sk Region. In the Irkutsk Region it runs the Bratsk pulp and containerboard mill and the Ust-Ilimsk pulp mill. In summer 2007 Ilim Pulp Enterprise was reorganized and renamed Ilim Group.[4] The company has strengthened its position in the Russian forest sector considerably. In August 2007 Ilim Group and International Paper decided to establish a new joint venture. According to press reports, International Paper has bought 50 per cent of the shares of Ilim Holding S.A. Through the joint venture, the companies seek to strengthen their position in the pulp and paper business in Russia (Grishkovets 2007).

The Arkhangel'sk pulp and paper mill is located in the city of Novodvinsk in the Arkhangel'sk Region. The company produces pulp, paper, board and fibreboard. The Arkhangel'sk mill was turned into a joint-stock company in 1992. Until 2003 it was part of a huge timber holding company, Titan Group. Since 2003 the Austrian company Pulp Mill Holding has functioned as the managing company of the mill. The Arkhangel'sk mill controls several logging enterprises and OAO Arhbum, the mill's exclusive supplier (Arkhangel'sk pulp and paper mill). In recent years, Pulp Mill Holding and Continental Management have struggled for control of the Arkhangel'sk mill.

Continental Management Timber Industrial Company is also a significant actor in the Russian pulp and paper industry. It is controlled by Oleg Deripaska's company, Basic Element, known until 2001 as Siberian Aluminium (SibAl).[5] According to Continental Management's web pages the company was established in March 2002 pursuant to a decision by Basic Element to develop a lumber-related line of business. In Northwest Russia, Continental Management operates

---

4   I use both names depending on the date involved.
5   I use both names depending on the date involved.

in the sawmill and wood-processing industries (Continental Management Timber Industrial Company). Continental Management has been seeking to expand its business networks vigorously. The company has generated a lively debate among Russian pulp and paper producers, since it has attempted to take over some large pulp and paper mills. Continental Management and Ilim Pulp Enterprise struggled for control over the Kotlas, Bratsk and Ust-Ilimsk combines for several years. Recently, Continental Management has attempted to seize control of the Arkhangel'sk pulp and paper mill and OAO Volga.

Large foreign forest enterprises have also shown growing interest in the pulp and paper business in Northwest Russia and some Russian pulp and paper combines are already controlled by them. Russia's huge natural resources and growing markets offer enormous opportunities, but these opportunities entail huge risks (Ernst&Young 2007). For instance, the Swedish company AssiDomän failed in its attempt to run the Segezha mill. (See more in chapter 8.) Nonetheless, some foreign direct investments have been more successful. The Svetogorsk pulp and paper mill in the Leningrad Region and the Syktyvkar pulp and paper mill in the Republic of Komi are among the most successful pulp and paper combines in Russia and are both controlled by foreign companies. The Syktyvkar mill is controlled by an Austrian company, Neusiedler, which is part of Mondi Europe, owned in turn by the company Anglo-American. International Paper bought the Svetogorsk combine already in 1998. As already mentioned, International Paper cooperates with Ilim Group as well.

Ownership Disputes Over Russian Pulp and Paper Mills

In the Russian media the ownership disputes over pulp and paper mills have been called 'forest wars'. According to the newspaper *Kommersant*, the starting point of these 'wars' was the struggle between Continental Invest and its partner Energoprom for control over Ust-Ilimsk Forest Industry Complex (Butrin 2004a). When Energoprom had seized control of the complex, the management of Continental Invest asked SibAl for help. Under the auspices of SibAl, Continental Invest managed to win back control over the Ust-Ilimsk combine. However, SibAl swallowed up Continental Invest and in this way acquired a new resource: managers with an in-depth knowledge of the forestry business. The group then started to look for other target enterprises and its gaze fell on the neighbouring Bratsk Forestry Plant, which belonged to Ilim Pulp Enterprise. SibAl acquired the plant by means of a bankruptcy procedure. However, the change in ownership was only temporary and eventually Ilim Pulp Enterprise managed to acquire control over Bratsk Forestry Plant and bought the Ust-Ilimsk combine as well (Pronin 2003; Klimenko 2003).

The struggle for the ownership of Ust-Ilimsk Forest Industry Complex had far-reaching consequences and in many respects determined the development of the entire sector (Butrin 2004a). As a result of the conflict, Oleg Deripaska's company

Siberian Aluminium (SibAl) entered the Russian forest sector and, with its allies, launched a real takeover wave. In 2001 the group announced their plan to move a significant amount of capital from the aluminium, oil, and banking sectors to the forest industry, in particular pulp and paper production, in order to establish a large timber holding company. This meant a serious takeover threat, since the main assets of Ilim Pulp Enterprise and the Arkhangel'sk pulp and paper mill were considered to be the basis of the new holding company (Butrin 2004a ; 2004c).

Basic Element (the former SibAl) established Continental Management in 2002 using the assets of Continental Invest. A new conflict between Continental Management and Ilim Pulp Enterprise began in spring 2002, when Continental Management attempted to seize control of the Kotlas and Bratsk combines (Klimenko 2003). During the takeover attempt, Continental Management acquired 61 per cent of the shares of the Kotlas combine and 83 per cent of the shares of the Bratsk combine (Sikamova 2004a). In summer 2002 the newspaper *Vedomosti* reported that Ilim Pulp Enterprise was going to lose the Kotlas mill as Continental Management had started to cooperate actively with the combine's clients (Novolodskaya et al. 2002). Nonetheless, Ilim Pulp Enterprise refused to give up control over its major combines. Moreover, the employees of the Kotlas combine supported Ilim Pulp Enterprise. They regarded Continental Management as an intruder and the newspapers reported that there was even a threat of armed conflict between the opposing parties (Bratkov 2002; Volkov V. 2004). The ownership disputes over the Kotlas and Bratsk combines lasted until 2004, when the parties eventually managed to reach a settlement. The end of the lengthy ownership disputes was regarded as a positive signal and the agreement between Ilim Pulp Enterprise, Continental Management and its partner Promyshlenno-stroitel'nyi Bank was reported widely by the Russian press. Ilim Pulp Enterprise retained control over the Kotlas and Bratsk combines. According to newspaper reports, Promyshlenno-stroitel'nyi Bank became a shareholder of the companies belonging to Ilim Pulp Enterprise and gave up its shares in Continental Management. At that time, the bank owned 20 per cent of the shares of the Arkhangel'sk mill and, pursuant to the agreement, sold them to Continental Management (Sikamova 2004a).

The sale of these shares to Continental Management was a significant turn of events and soon a new ownership dispute started in the Arkhangel'sk Region. Newspaper reports pointed out that the management of the Arkhangel'sk pulp and paper mill was very aware of the hostile takeover threat (Sikamova 2004c; Butrin 2004c). The Arkhangel'sk mill was controlled by Titan Group until autumn 2003. In summer 2003 an Austrian company, Pulp Mill Holding, one of the foreign shareholders of the mill, obtained permission from Russian antimonopoly officials to consolidate 75 per cent of the company shares. Some specialists suggested that the purpose of this friendly takeover was to protect the company against hostile takeover attempts (Fëdorov 2003). According to press reports, Pulp Mill Holding managed to consolidate only 65 per cent of the shares of the Arkhangel'sk mill. The transaction was contested by a minority shareholder of the mill – Hover Group. As a result of Hover Group's lawsuit, the Moscow Arbitration Court held that the consolidation of the shares of Arkhangel'sk pulp and paper mill was illegal. The press considered this decision a serious warning signal of a hostile takeover (Butrin 2004c). Moreover, it was suggested that Continental Management was behind the lawsuit, but Continental Management categorically denied this (Sikamova and Shcherbakova 2004).

Continental Management has managed to increase its share ownership in the Arkhangel'sk pulp and paper mill. As mentioned above, on the basis of the agreement between Ilim Pulp Enterprise, Continental Management and Promyshlenno-stroitel'nyi Bank, Continental Management acquired 20 per cent of the shares of the Arkhangel'sk mill. Furthermore, it bought 12.5 per cent of the shares of the mill from Wilfried Heinzel (Khrennikov et al. 2004). Continental Management has actively participated in the governance of Arkhangel'sk pulp and paper mill. There is a deep mistrust between Continental Management and Pulp Mill Holding. The conflicting shareholders have not managed to reach an understanding on how the company should be run.

In addition to the above-mentioned corporate conflicts, the struggles for the ownership of the Sokol pulp and paper mill and OAO Volga have attracted the interest of the Russian media. The Sokol mill is located in the Vologda Region. Vektor and Gruppa Fox struggled aggressively for control over the enterprise. Newspapers reported that there was even an armed conflict between the parties in the autumn of 2004 and some people were injured in a clash with the armed troops. In the press the opposing parties accused each other of resorting to violence (*Pravda Severa* 2004; Sikamova 2004b). Nowadays, it seems that the situation at the Sokol mill is more peaceful. The mill is a part of the newly established holding company, Investlesprom (Lesgazeta.ru 2007). OAO Volga, on the other hand, is in the middle of an ownership dispute. OAO Volga is located in the Nizhniy Novgorod Region and is controlled by Ost-Vest Group. According to press reports, Continental Management offered to buy OAO Volga already in 2004. However, Ost-Vest Group refused to sell the company, because it planned to build a holding company based around OAO Volga (Khrennikov and Simakov 2005). Ever since, OAO Volga has been the subject of numerous inspections and audits carried out by local and state authorities. The company management has been accused of tax evasion. It has been suggested that all these difficulties are connected with the takeover attempt (Chichurina 2005).

Enterprise Takeovers

Takeovers, their motives and consequences, and whether the takeover market operates efficiently have been the subject of numerous theoretical and empirical studies. The existing research is extensive and it is beyond the scope of this chapter to review it in detail. An enterprise takeover is more than just a transfer of ownership. Typically the firm acquired undergoes a major reorganization. The question whether these changes create or destroy value has caused considerable controversy among business people and scholars. Those threatened by the changes that restructuring brings about argue that takeovers are damaging to the economy, harmful to the morale and productivity of organizations and pressure executives to manage for the short term. In other words, enterprise takeovers destroy rather than create value (Jensen 1988). However, others argue that takeovers are motivated by value improvements. Takeovers represent productive entrepreneurial activity that improves control and management of assets and helps put them to more productive uses. Changing technology or market conditions require the restructuring of corporate assets. On the other hand, some takeovers occur because the incumbent managers are incompetent (Jensen 1986, 1988). Enterprise takeovers reduce agency costs stemming from the separation of ownership and control. The company's efficiency is increased by replacing managers who are either less competent or do not act in the shareholders' best interest (Burkart et al. 2006).

Enterprise takeovers are classified as friendly or hostile depending on managers' and shareholders' attitudes towards the takeover bid. A hostile takeover is an attempt to gain control over the financial and economic operations or assets of the target company in the face of resistance from the target company management and shareholders (Radygin et al. 2003). However, it has been suggested that this simple distinction is not as clear as it seems. Transactions may contain both hostile and friendly elements and many voluntary transactions would not occur without the threat of a hostile takeover (Jensen 1988; Schwert 2000). While the hostile takeover is a well-known concept in Russia, some Russian scholars prefer to use the term 'redistribution of ownership' instead of 'merger and acquisitions' or 'takeover' in order to underline the characteristic features of Russian enterprise takeovers (e.g. Radygin et al. 2003; Belyaeva and Belyaev 2005). Apparently, some authors suggest that abuse of legislation and corruption are inextricably connected with hostile takeovers in Russian. Nonetheless, it has also been highlighted that takeovers can be carried out in a legal manner and in such cases can increase the efficiency of the acquired firm (Demidova 2007; Chuyasov 2007; Pozdnyakov 2006).

Russian scholars tend to argue rather unanimously that hostile takeovers became a prevalent phenomenon after the 1998 financial crisis. The laws regulating mergers and acquisitions were new and unfamiliar to company managers and shareholders as well as judges and the authorities responsible for the securities market. Moreover, the securities market was rather underdeveloped. Under such circumstances, hostile takeovers were regarded as the cheapest and simplest way to acquire control over an attractive company. In big cities like Moscow, hostile enterprise takeovers were also used as a device for acquiring real property. In those cases, the new owners had no long-term business interest and the enterprises acquired were eventually liquidated (Belyaeva et al. 2005). Although this contribution deals merely with takeover attempts in the Russian pulp and paper industry, it should be noted that ownership disputes are not an unknown phenomenon in other industrial sectors. According to Vladim Volkov, since 1998 the principal targets of enterprise takeovers in Russia have been:

a) lucrative export-oriented enterprises such as aluminium, steel or cellulose production and electric machine-building; b) enterprises of the so-called 'fuel and energy complex'; c) ore-processing plants and enterprises that supply vital raw materials to metallurgical enterprises; d) enterprises in consumer industries such as food, alcohol and cosmetics; and e) any enterprises possessing valuable assets that can profitably be sold (Volkov V. 2004).

Mergers and acquisitions are regulated by the law on joint-stock companies. Provisions dealing with mergers and acquisitions can be found in the Russian Civil Code as well. The law on joint-stock companies regulates the acquisition of shares in a company and the procedures that have to be followed during the transaction. Other laws, such as antimonopoly regulations, must be taken into account as well. The protection set forth in legislation is designed to ensure that the shareholders of any company to be acquired receive fair market value for their investment, even if the company is acquired in a hostile takeover. However, in practice the regulation on mergers and acquisitions has been superseded by powerful informal practices. The institution of bankruptcy and the minority shareholders' rights as regulated by the law on joint-stock companies have often been used as takeover instruments in the enterprise takeovers in the Russian forest sector. It is commonly acknowledged that this is a serious problem (Khakimov 2004; Volkov V. 2004; Markov 2007). Later I will review in more detail the possibilities that the laws on insolvency and on joint-stock companies have offered to companies planning takeovers. Groups planning takeovers constantly search for new takeover methods, as old ones become too risky. Accordingly, the laws on insolvency and on joint-stock companies are not the only ones that have been abused; for instance, criminal legislation has been used as a takeover tool as well (Loginov et al. 2007).

The abuse of legislation in hostile takeovers illustrates that borrowing legal rules from one legal system to another is a challenging task. Theories of legal transplants, such as Legrand's, which focus excessively on the legal system and legal culture and not its interplay with the rest of the society, do not offer tools to study law in transition. Theories like Watson's, which regard law as merely a lawyers' playground and suggest that law evolves separately from society, also have rather little to say where Russian law is concerned. Douglass C. North, on the other hand, examines institutional change in a wider perspective (see Chapter 2 of this volume). The performance of legal rules depends upon the overall environment in which they are applied. According to North, formal rules, like laws, can be altered relatively quickly, but informal constraints such as customs and codes of conduct change more slowly. In Russia the transition to a market economy led to the destruction of the old formal institutional framework and the emergence of new formal rules. However, many informal constraints have survived despite the sweeping changes. According to North, it is the interplay between formal and informal constraints, as well as the type and effectiveness of enforcement, that shapes the rules of the game in a society (North 1990, 1997).

The Russian laws on insolvency and joint-stock companies have been drafted in accordance with Western standards. However, legal rules which work well in one country might bring unexpected outcomes when transferred to another. Transplanted laws confront different social economic and institutional contexts. This 'transplant effect', that is, the mismatch between pre-existing conditions, institutions and a transplanted law, weakens the effectiveness of the transplanted law (Berkowitz et al. 2003). As Teubner puts it, a foreign rule is an irritant, or 'outside noise', in a legal system. It not only irritates the minds of lawyers who apply the rule but disturbs the law's binding arrangements as well. A foreign rule triggers a whole series of new and unexpected events in the environment into which it is transferred. Moreover, the rule itself undergoes transformation and does not continue to play its old role when applied in new circumstances (Teubner 1998). It seems that even the Russians have been rather surprised at the way in which the laws on insolvency and joint-stock companies have functioned in the country and at the role of these laws in business conflicts. I suggest that the underlying reasons for the abuse of legislation in hostile takeovers are corruption and the involvement of state and local authorities in business conflicts. In these kinds of circumstances, the transplanted law inevitably undergoes transformation. The actions of economic subjects depend on the means available to them and legal rules become means rather than constraints if economic subjects have a strong influence on the interpretation and enforcement of laws (Volkov V. 2004).

It has been suggested that the characteristic feature of hostile takeovers in Russia is extensive use of what are known as 'administrative resources' (Demidova 2007; Volkov V. 2004). This term refers to the capability of groups planning takeovers to assemble and coordinate a network of actors who are powerful enough to support the acquisition of assets. Often, the actors in such a network belong to the state and are mobilized by the group planning a takeover through a combination of formal and informal methods. The use of 'administrative resources' allows a group planning a takeover to achieve at least three main objectives: (1) to assure that the law is applied in favour of the takeover, (2) to ensure security during the takeover process, and (3) to create insecurity and exert pressure on the opponent (Volkov V. 2004). Below I will review in more detail the role of administrative resources in the hostile takeovers in the Russian forest sector.

The courts have played a key role in hostile takeovers in Russia. Groups planning takeovers have taken advantage of corrupt judges and the defects of the Russian court system. The central role of corruption in enterprise takeovers is widely acknowledged (Volkov V. 2004; Chuyasov 2007). Groups planning a takeover often use their informal connections to ensure that the law is applied in favour of the takeover attempt and that court decisions are enforced quickly. The jurisdiction of Russian courts is rather unclear, which makes the court system vulnerable. Corporate conflicts have been heard by both arbitration courts and courts of general jurisdiction. There are also instances where the same case has been heard by two or more different courts. It seems that parties planning takeovers find that the courts of general jurisdiction serve their purposes better than the arbitration courts. In practice corporate conflicts have been disguised as disputes concerning labour issues, for instance, because such disputes can be heard by the courts of general jurisdiction (Protskuryakova 2006). Another serious problem is that lawsuits connected with takeover battles are frequently brought in courts that are located as far as possible from the target company's registered office. Groups planning takeovers often seek to bring the suits in courts where they are more likely to obtain decisions supporting their takeover attempt (Protskuryakova 2006). This hinders the owners of the target company in defending their rights efficiently, as the defendant does not necessarily receive sufficient information on the proceedings. Travelling can be difficult and different time zones have to be taken into account as well (Chuyasov 2007).

Characteristics of Ownership Disputes in the Russian Pulp and Paper Industry

The ownership disputes over Russian pulp and paper mills are classified as hostile enterprise takeovers by the Russian press. The management of the mills, as well as shareholders and employees, have vigorously resisted takeover attempts and in extreme cases there have been even armed conflicts between competing shareholders. It seems that the struggles for ownership of Russian pulp and paper mills are to some extent connected with the privatization of these enterprises. Ownership arrangements based on the privatization of enterprises have proven to be rather vulnerable. For instance, in the takeover battles for the Kotlas, Bratsk and Arkhangel'sk combines and for OAO Volga the decisions and choices made during the privatization of the mills were contested in order to alter the structure of their ownership.

The privatization of the Kotlas pulp and paper mill was contested by a minority shareholder of the mill. The shareholder filed a suit in the district court of Kemerovo and alleged that the majority shareholder of the combine, Ilim Pulp Enterprise, had failed to observe the privatization agreement of 1994 and to fulfil the investment plan prescribed by the agreement, causing the mill losses. The court ruled in favour of the plaintiff and ordered compensation from Ilim Pulp Enterprise equivalent to 60 per cent of the value of the shares of the Kotlas combine. Obviously, the main objective of the lawsuit was not to protect the interests of the company and its shareholders but to arrange the sale of shares of the mill to the party planning the takeover. One outcome of the court decision was that the shares of the Kotlas mill were confiscated and eventually sold to Continental Management's front company. A sale of the shares of the Bratsk combine was arranged in a similar manner (Volkov V. 2004; Yeremenko 2004).

During the takeover battle, the privatization of the Arkhangel'sk pulp and paper mill and particularly the establishment of a holding company, Severnaya Tsellyuloza, have been contested. Severnaya Tsellyuloza was founded in 1994. At that time, 20 per cent of the shares of the Arkhangel'sk mill were owned by the Arkhangel'sk Region. When the mill was privatized, these shares were transferred to the share capital of Severnaya Tsellyuloza. During takeover battle, this transaction and the privatization of the mill have become the object of a lengthy criminal investigation (Rebov 2005). The press reported that at the end of 2005 16 per cent of the shares of the Arkhangel'sk pulp and paper mill were confiscated on the grounds of this criminal investigation. The confiscated shares belong only to Pulp Mill Holding. The press claimed that it was evident that the criminal investigation and the confiscation of the shares are closely connected with the struggle between Pulp Mill Holding and Continental Management for control of the mill (*Pravda Severa* 2007; Temkin 2006, 2005b).

Finally, there has also been controversy surrounding the privatization of OAO Volga. In 2005 local officials of the Nizhniy Novgorod Region responsible for property issues brought a suit against OAO Volga in arbitration court alleging that the decisions made during the privatization of the enterprise in 1994 were null and void. However, according to press reports, the court ruled in favour of OAO Volga. The main argument of the court was that the period for filing the suit had expired and the plaintiff thus did not have the right to sue (Fil'tsov and Milyaev 2005). The representatives of the mill considered the motive behind the lawsuit to be a hostile takeover (Regnum 2005).

The Russian media, as well as the state and local authorities, have played a significant role in the ownership disputes in the pulp and paper industry. The media have reported on the takeover battles in the forest sector in detail, and opposing parties have taken advantage of the media's interest by mounting campaigns to promote their own points of view and put pressure on their rivals. The news dealing with the ownership struggle between Ilim Pulp Enterprise and Continental Management illustrates this well (Khrennikov 2003). The ownership struggles for the Arkhangel'sk pulp and paper mill and OAO Volga show that in takeover battles both parties have taken advantage of their good relations with the state and local authorities. What are known as 'ordered inspections', carried out by state officials such as the tax authorities, have been a common phenomenon in the takeover battles. According to press reports, both the Arkhangel'sk mill and OAO Volga have become the targets of numerous inspections and audits carried out by different state authorities (Kistanov 2006; *Pravda Severa* 2006). For instance, the representatives of OAO Volga have said that the number of inspections and audits increased substantially after the management of the combine refused to sell the company to Continental Management (Khrennikov and Simakov 2005). Groups planning takeovers have used the media as a device for mobilizing the state and local authorities. Typically, a group publishes an announcement in a newspaper suggesting, for example, that the target company has substantial tax arrears. The announcement then attracts the interest of the tax authorities and they start a tax audit. Groups planning takeovers have also taken advantage of their connections to the State Duma. Some members of the Duma have helped these groups by persuading other state authorities to initiate different kinds of inspections and audits (Kistanov 2006).

Numerous inspections and audits cause extra work and harm to the everyday operations of the target enterprise. Inspections and audits have also been used as a device for collecting information on the target enterprise. A group planning a takeover needs information on the target company's management, financial status, ownership structure and the like. Collecting information can be a hard task, since Russian companies are often not very transparent. In practice, there have been several instances where state officials, such as the tax authorities, have acted on behalf of groups planning takeovers and collected information on the target enterprise for them (Khakimov 2004).

On the other hand, takeover targets have also used their 'administrative resources'. The Arkhangel'sk mill and OAO Volga have started to work together against hostile takeover attempts. Such an alliance of large enterprises is not so common in Russia. Representatives of both companies have spoken rather frankly in the press about the kind of difficulties their companies have met with. The management of OAO Volga has also worked with the state and local authorities in order to defend the company against a takeover attempt. Both enterprises have actively participated in the work of different state committees. For instance, they have worked together in a special committee whose objective is to monitor hostile takeovers and to find solutions to the problems caused by ownership struggles. The committee was established by the Russian Chamber of Commerce and Industry (*Rossiyskaya Gazeta* 2006). The Russian Association of Organizations and Enterprises of the Pulp and Paper Industry (*RAO Bumprom*) has also openly supported these companies as they have campaigned against hostile takeovers.

Hostile Enterprise Takeovers Behind Bankruptcy Proceedings

The 1998 law on insolvency was regarded as a great step towards more efficient and better-functioning enterprises. The law was drafted according to western standards. The previous law, which dated from 1992, had proven ineffective in practice (Tompson 2003). While the number of bankruptcies in fact skyrocketed after the law on insolvency came into force in 1998, the main reason for this development was that bankruptcy was used as a device for redistributing ownership rights and control over enterprises (Volkov V. 2004). Bankruptcies became a profitable business and a number of firms and groups specializing in 'ordered bankruptcies' emerged. Usually lawyers with comprehensive knowledge of the nuances of insolvency legislation constituted the core of the group. Former judges, officials from law enforcement agencies and other state bodies directly related to economic regulation participated in the bankruptcy business as well. The primary task of these firms and groups was to chase down appropriate takeover targets for their clients and help their clients during bankruptcy proceedings (Chernigovskiy 2001).

The 1998 law on insolvency had several shortcomings that made it an effective tool for engineering a change of management in target enterprises. Moreover, enforcement of the law was inefficient and corruption played an essential role in 'ordered bankruptcies'. Bankruptcy proceedings could be initiated rather easily even by minor creditors. The law permitted arbitration courts to initiate bankruptcy procedures against an enterprise whose outstanding debt exceeded the equivalent value of 500 monthly minimum wages[6] and if the debt was not repaid within three months of the due date (Volkov V. 2004). Thus, one prerequisite for using bankruptcy as a takeover tool was that the target company owed some money to the party planning a takeover. In practice a group planning a takeover often persuaded creditors to collaborate with it or simply bought the debts of the target company in the form of a bill of exchange (Radygin et al. 2005). Russian legislation provides that the debtor should be informed about a change of creditor. However, in practice the debtor was given no notification at all, since a group planning a takeover does not want to warn the target company beforehand. Instead of being sent the notification required by law, the target enterprise received, in a registered letter, a blank sheet of paper or some advertising material. Later on, the demand for payment was sent to the target enterprise in a similar manner. Finally, in the court proceedings, postal receipts and copies of the documents presumably contained in the letters thus mailed served as evidence that the debtor had failed to repay its debt on time. In practice, rather dubious methods were sometimes used in order to hinder the target company from repaying its debts. For instance, debtors were sent incorrect bank details (Chernigovskiy 2001; Radygin et al. 2005).

One of the principal shortcomings of the 1998 law on insolvency was that the functions of temporary and external managers were supervised inefficiently. The main task of the managers was to balance the interests of the debtor and the creditors and to ensure that the rights of all parties were respected (Tompson 2003). However, in practice the managers merely attended to the interests of the parties planning takeovers. Temporary managers were appointed by arbitration courts but very often the court decisions were prearranged and representatives of the parties planning takeovers were nominated as temporary mangers (Volkov V. 2004). Temporary managers had to ensure, among other things, that a bankruptcy had not been 'ordered' but rather often they ignored even the most obvious signs of such a bankruptcy.[7] In addition, it was rather common that temporary managers manipulated the debts of the target company in order to ensure that the bankruptcy proceedings took the course desired by the groups planning the takeover (Radygin et al. 2005).

As bankruptcy proceedings progressed, creditors appointed external managers, who replaced the management team of the bankrupt company. They had wide powers and it was during the external managers' administration that the enterprise

---

6   In Russia, the minimum wage is regulated by a federal law. Provisions dealing with salaries can also be found in the Russian Labour Code.

7   Radygin et al. (2005) describe a bankruptcy case where a temporary manager confirmed that there were no signs of a premeditated bankruptcy. However, it later became clear that the debtor had been willing to repay the debt but the creditor had purposely hindered the payment. The creditor had given the debtor incorrect bank details and the debtor was therefore unable to pay the debt on time.

takeover occurred (Radygin et al. 2005). The arbitration court had to confirm the appointment of an external manager on the basis of the minutes of a creditors' meeting and in practice groups planning takeovers could rather easily influence the court decision. Temporary managers were responsible for taking the minutes of the creditors' meeting, but no one ensured that the decisions of the meeting were actually written into the record. There were instances where the arbitration court appointed a person as an external manager against the will of the creditors because the temporary manager had manipulated the minutes of the creditors' meeting (Volkov A. et al. 1999).

The administrative resources of groups planning takeovers played an essential role in such takeover schemes. The groups had to ensure that their representatives were appointed as temporary and external managers. Efficient proceedings and the rapid enforcement of judgments were also essential for a successful takeover. Rather often special police forces and other armed troops were used to ensure the change of management at the target enterprise. When a new management team had started its work, it proved to be very difficult in practice for the old management to reverse the situation (Volkov V. 2004). The outcome of a takeover depended on the objectives of the aggressor. In some cases, the new management was only interested in quick, short-term profits; in others, the new management had a long-term business interest and the acquired company remained under the control of the new team (Volkov A. et al. 1999; Volkov V. 2004).

Bankruptcy proceedings were used as a takeover instrument in the metallurgy, oil and gas industries as well as the pulp and paper industry. For instance, SibAl acquired control over the woodworking plant Bratsk Complex Holding by means of a bankruptcy procedure. By the end of the 2001, the electricity debt of Bratsk Complex Holding to the local energy company Irkutskenergo was 750 million roubles. Moreover, in the same year SibAl had acquired 30 per cent of the shares of Irkutskenergo. Bratsk Complex Holding offered to repay the energy debt in order to avoid bankruptcy. However, SibAl managed to engineer a change of management at Bratsk Complex Holding by means of a minority shareholder's lawsuit, as a result of which the court dismissed the incumbent management of the company. The court decision was enforced rapidly. A new management team started its work with the help of court bailiffs and armed guards and it immediately cancelled the repayment agreement with Irkutskenergo. Accordingly, Irkutskenergo, which was controlled by SibAl, could initiate bankruptcy proceedings; SibAl seized control of Bratsk Complex Holding but only temporarily. Eventually, the opposing parties reached a settlement and Ilim Pulp Enterprise retained control over Bratsk Complex Holding (Volkov V. 2004; Pronin 2003).

The practice of using bankruptcy as a hostile takeover tool generated harsh criticism in Russia, leading to the adoption in 2002 of a new insolvency law. The new law and closer supervision of its application have considerably reduced the abuse of bankruptcy in hostile takeovers (Volkov V. 2004; Belyaeva et al. 2005).

The Russian Law on Joint-Stock Companies and Hostile Enterprise Takeovers

Clearly, the objective of the minority shareholders' rights as regulated by the Russian law on joint-stock companies is to provide minority shareholders a possibility to influence the governance of the company and to protect them against the misconduct of controlling shareholders. The hostile takeover attempts in the Russia pulp and paper industry illustrate that the minority can sometimes oppress the majority. According to Volkov, the abuse of minority shareholders' rights in takeover battles was a rather rare practice before 2001. It became more common through attempts to change the ownership of the major enterprises in the pulp and paper industry (Volkov V. 2004).

According to press reports, minority shareholders' lawsuits have been rather common in ownership disputes over pulp and paper mills. For instance, several lawsuits have been filed against the Arkhangel'sk pulp and paper mill by minority shareholders. In August 2005, a minority shareholder, Watech Ltd, brought a suit against the mill in the arbitration court of the Arkhangel'sk Region. Watech Ltd alleged that the contract between the mill and its exclusive distributor, Arhbum, was disadvantageous to the mill. According to *Vedomosti*, Watech Ltd was strongly supported by Continental Management and the lawsuit was regarded as being closely connected with the ongoing ownership dispute (Temkin 2005a). In the same year there was a controversy between Watech Ltd and Arkhangel'sk pulp and paper mill over a minority shareholder's right to get information on the financial position of the mill (Rebov 2005). Furthermore, a minority shareholder's lawsuit played a central role in the struggle for control over Bratsk Complex in 2001. As already mentioned, the company's old management team was dismissed as a result of the minority shareholder's appeal and the new management helped the attacking party to initiate bankruptcy proceedings. Minority shareholders' suits have also been used to arrange the sale of shares in the target company. As noted above, minority shareholders' lawsuits were the starting point of the controversy over the ownership of Kotlas and Bratsk combines in 2002. The privatization of the mills was contested by minority shareholders and as a result of the lawsuit a significant block of shares in the mills was sold.

For a group planning a takeover to be able to use minority shareholders' rights, it has to own a sufficient number of shares in the target company. This can be a difficult aim if the target company is controlled by company insiders such as employees and managers and they are not willing to sell their shares. Nonetheless, it is possible to take advantage of minority shareholders' rights without share ownership if minority shareholders agree to collaborate. If there are conflicts of interest between minority shareholders and controlling shareholders it can sometimes be rather easy to persuade minority shareholders to contribute to a hostile enterprise takeover (Klyuchko 2003). In practice, there have also been cases in which the minority shareholder has been purely fictional. In other words, it has turned out that the shareholder filing the lawsuit did not actually own any shares at all (Volkov V. 2004).

Rather often minority shareholders' right to sue has been abused in order to obtain a court order for precautionary measures. Clearly, the main motive of these minority shareholders' appeals is to obtain a court order restricting the rights of the majority shareholder (Volkov V. 2004). Thus, the protection of minority shareholders' interests is usually not essential for the plaintiff. Often the court orders are based on a minority shareholder's claim that some acts of the majority shareholder have caused the minority shareholder damage and that the appropriate way to protect the minority shareholder's interest is to enjoin the majority shareholder from voting at shareholders' meetings (Klyuchko 2003; Chuyasov 2007). In practice, court orders have been used to confiscate the shares belonging to majority shareholders. They have also prohibited the target company's governing bodies and registrar, who is responsible for keeping the register of shareholders, from making certain decisions or carrying out their duties (Romanenko 2007). In practice, precautionary measures have proven to be a useful device for engineering decision-making in shareholders' meetings. As a rule, decisions at shareholders' meeting are taken by a simple majority, an exception being decisions for which the law requires a qualified majority of three-quarters. When the majority shareholder is prohibited from voting at the shareholders' meeting by a court order, the minority shareholder can control the decision-making.

Since a change of the management team is essential for the success of any hostile takeover attempt, the boards of directors of the target companies have played a central role in takeover battles. It is crucially important for a group seeking to take over another company to have its own representatives elected as members of the governing bodies of the target company. In Russian joint-stock companies, the members of the board of directors are elected by the shareholders' meeting. Accordingly, in practice the meetings have been major battlefields in ownership disputes. According to the law, any shareholder who owns no less than 10 per cent of shares with voting rights has a right to ask the board of directors to call an extraordinary shareholders' meeting. The law on joint-stock companies protects shareholders. The discretionary power of the board of directors is very limited and therefore the board often have no choice but to arrange this meeting (Klyuchko 2003). The right of minority shareholders to call an extraordinary shareholders' meeting without the support of the board of directors has been abused by groups planning takeovers. According to the law, shareholders are permitted to call an extraordinary shareholders' meeting by themselves if the board of directors refuses to arrange extraordinary shareholders' meeting without sufficient reason[8] or if the board fails to make its decision without delay. In these cases, shareholders have rather free hands to organize the meeting as they wish.

---

8   The circumstances under which the board of directors can reject the request of shareholder are regulated exhaustively by the Russian law on joint-stock companies.

They can decide the date and the place[9] of the meeting as well as the agenda (Klyuchko 2003). In hostile takeovers, minority shareholders' requests have often deliberately been formulated such that the board of directors of the target company cannot help refusing to call an extraordinary shareholders meeting. Such a decision then gives the shareholders planning a takeover an opportunity to organize the extraordinary shareholders' meeting as they please (Romanenko 2007). Often the majority shareholder is prohibited from participating in the shareholders' meeting arranged by the shareholder planning a takeover. For instance, a court order for precautionary measures is often abused in order to make sure that the shareholder planning a takeover can control decision-making at the shareholders' meeting.

Shareholders' competition for the seats on the board of directors has led to situations where several shareholders' meetings have been organized one after another by opposing parties. For instance, in the struggle for ownership of the Kotlas and Bratsk mills, Continental Management and Ilim Pulp Enterprise fought for seats on the boards of directors of the companies. Both parties organized their own shareholders' meetings, the main objective of which was to elect the members of the boards of directors. As a result, both the Kotlas and Bratsk combines had two boards of directors in the summer of 2002. Obviously, there was a serious struggle between Continental Management and Ilim Pulp Enterprise over whose board of directors was the more competent (Khrennikov 2003; Temkin 2004). Undoubtedly, having parallel management teams causes serious difficulties for any company. The situation at the Kotlas and Bratsk mills in 2002 was also uneasy because, in addition to having two competing management teams, the companies had two shareholders' registers as well (Khrennikov 2003; Temkin 2004). In 2004, for instance, the arbitration court of the Arkhangel'sk Region dealt with the ownership struggle for the Kotlas mill.[10] The proceedings mainly concerned the validity of the decisions made at the shareholders' meetings and which of the two registrars should be regarded as the legal registrar of the mill. The company had two shareholders' registers, since it had two boards of directors and two general managers. One management team had changed the registrar of the company but the competing management team did not recognize the change. The proceedings in the arbitration court illustrate well the complex situation at the Kotlas mill in 2004. The mill was represented in court by both competing management teams. The representatives had very different opinions about how the dispute should be resolved. The ownership dispute over the mill generated several sets of court proceedings but the courts could not end the takeover battle. The ownership dispute between Continental Management and Ilim Pulp Enterprise was finally settled when the conflicting parties managed to negotiate an agreement.

---

9   Naturally, a company's byelaws may contain provisions concerning the location of shareholders' meetings.

10   Arbitration court of Arkhangel'sk Region, decision no. 05-12378/02-588/17 and no. 05-9374/03-17

Concluding Remarks

Enterprise takeover attempts in the Russian pulp and paper industry have resulted in lengthy struggles for the ownership of the mills. Naturally, Russian pulp and paper producers, as well as RAO Bumprom, have been very concerned about the negative impacts of hostile enterprise takeovers on the investment climate in the Russian forest sector (*Rossiyskaya Gazeta* 2002). Hostile enterprise takeovers involving abuse of legislation have also generated a lively debate on the possibilities of restricting such takeover activities. It seems that new laws are seen as one solution to the problems caused by takeovers. Accordingly, since the end of 2006 the Ministry of Economic Development and Trade and the State Duma have worked on a law project whose objective is to modify several laws in order to restrain abuse of legislation in hostile enterprise takeovers (Protskuryakova 2006).

Clearly, the legislator is trying to fill in the loopholes in the laws that have been used as takeover tools. The same approach was also used when the law on insolvency was amended. The widespread practice of using bankruptcy as a takeover tool was the main reason why the law on insolvency was amended and therefore special attention was paid to the rules that in practice had functioned as takeover instruments (Radygin et al. 2005). While the 2002 law on insolvency has successfully reduced the role of bankruptcy in hostile takeovers, one should note that in the same year the law on joint-stock companies was amended and minority shareholders' rights were strengthened. Groups planning takeovers tend to look for new takeover methods as old ones become too risky. Perhaps they have discovered that the minority shareholders' rights regulated by the law on joint-stock companies serve their objectives better than the law on insolvency.

The law project focuses on the defects of the Russian court system and Russian company legislation. According to the project proposal, corporate conflicts will be heard exclusively by the arbitration court of the target company's registered office. Moreover, it is suggested that the meaning of the concept 'corporate dispute' should be defined in detail by the law (Romanenko 2007). The rationale here is to make it more difficult to disguise corporate disputes as other kinds of conflicts (Sterkin 2007, Protskuryakova 2006). The amendments are seen as preventing groups planning takeovers bringing lawsuits in courts that would most probably support their purpose. The abuse of precautionary measures in hostile takeovers is regarded as a serious problem as well. Therefore, it is suggested that the prerequisites for deciding on precautionary measures should be tightened (Romanenko 2007). There are also plans to amend Russian company legislation. It is suggested that shareholders' right to call an extraordinary shareholders' meeting should be restricted in order to avoid shareholders organizing competing shareholders' meetings. As these parallel shareholders' meetings are often called in order to elect a new board of directors, it is thought that restricting shareholders' right to convene extraordinary shareholders' meetings will solve the problems caused by parallel management teams (Romanenko 2007).

Law reforms are undoubtedly important but they are challenging as well. Russia's move to a market economy made it a large-scale borrower of Western models. This is understandable, since there was an urgent need for laws that fit a market economy. The Russian law on joint-stock companies, for instance, does not differ considerably from its Western models. Nonetheless, the abuse of minority shareholders' rights in enterprise takeovers reveals that in practice the law has not functioned as expected. Evidently, the problem has less to do with the letter of the law than with the environment in which it is applied. It is a rather common complaint that in Russia the transplanted laws are only on books and that the enforcement of legislation is inefficient. Certainly, the Russian court system has several shortcomings but this is not the only source of the problems. Corruption is an important informal way to 'get things done' and it plays a central role in enterprise takeovers. The law becomes a weapon, since the economic actors have a significant influence on how laws are applied and enforced. These conditions are challenging for the transplanted laws and they might function in a totally unexpected way. Transferring legal institutions from one context into another is something very different form substituting a part of one machine with a part from another. It is not an easy task and, as Teubner puts it, borrowed legal rules need careful implantation and cultivation in the new environment (Teubner 1998).

Continental Management has been one of the most aggressive enterprises in the Russian pulp and paper industry. Nonetheless, its takeover attempts have not been successful. Perhaps the target enterprises have been more powerful than Continental Management expected. The managers and shareholders of the pulp and paper mills have fought back. There has been a great deal of debate on corruption in Russia. A review of Russian newspapers clearly shows that corruption is being criticized more and more openly. Since legislation changes and bribery becomes more expensive and risky, it is sometimes suggested that aggressive attacks will become less common and hostile takeover methods more subtle (Loginov 2007). It is difficult to predict whether the 'forest wars' will continue in Russia and, if they do, what forms ownership disputes will take. However, it is rather evident that the ownership of pulp and paper mills is becoming concentrated in fewer hands and that stronger enterprises will emerge in the forest sector.

References

Antanta Capital Investment bank (2006), Russia's Timber Industry an Area of Uncertainty, April 25. <http://www.antcm.ru/ftproot/Files/wood_250406eng.pdf>.

Arkhangelsk pulp and paper mill <http://www.appm.ru>.

Belyaeva, I.Yu and Belyaev, Yu.K. (2005) 'Rossiyskiy rynok slijaniy i pogloshcheniy: Èvolutsiya i perspektivy razvitiya', *Finansy i Kredit*, no. 26, September 19.

Berkowitz, Daniel, Pistor, Katharina and Richard, Jean-Francois (2003), 'The Transplant Effect', *The American Journal of Comparative Law*, vol. 51, No. 1. (Winter, 2003), 163–203.

Bratkov, Vitaly (2002), 'Epidemic of proletarian mutinies in the Arkhangelsk Region', Pravda.ru, July 10.

Burkart, Mike C. and Panunzi, Fausto (2006), 'Takeovers', *ECGI – Finance Working Paper*, No. 118/2006. Available at SSRN: <http://ssrn.com/abstract=884080>.

Butrin, Dimitry (2004a), 'Timber industry 2000–2004', *Kommersant*, November 11 <http://www.kommersant.com/tree.asp?rubric=3&node=43&doc_id=487624 >, all the article in Russian as of July 5, 2004.

Butrin, Dimitry (2004b), 'Timber industry 1991–2000', *Kommersant* 20.8.2004. <http://www.kommersant.com/tree.asp?rubric=3&node=43&doc_id=307741 >, all the article in Russian as of January 29, 2002.

Butrin, Dimitry (2004c), 'Lesnaya tselesoobraznost', *Gazeta.Ru-Kommentariy*, http://www.gazeta.ru/comments/2004/05/a_112010.shtml.

Chernigovskiy, Maksim (2001), 'Bankrotsvo Zakazyvali?', *Kommersant*, 103P, June 8.

Chichurina, Yekaterina (2005), 'Bitva za Volgu', *Vedomosti*, 240 (1521) December 21.

Chuyasov, Andrey (2007), 'Mekhanizmy nezakonnogo pogloshcheniya predpriyatiy', *Pravo i Ekonomika*, 3, March 30.

Continental Management Timber Industrial Company <http://www.lpkkm.ru>.

Demidova, E. (2007), 'Brazhdebnye pogloshcheniya i zashchita ot nikh v usloviakh korporativnogo rynka Rossii', *Voprosy Ekonomiki*, 4, April 25.

Dimitrev, Valeriy (2007), 'Vozrozhdenie lesnoiy Solombaly oboshlos' bez inostrannogo kapitala', *Pravda Severa*, September 26.

Dudarev, Grigory, Boltramovich, Sergey and Efemov, Dimitry (2002), *From Russian Forests to World Markets. A Competitive Analysis of the Northwest Russian Forest Cluster*. ETLA, Research Institute of the Finnish Economy (Helsinki: Taloustieto Oy).

Eronen, Jarmo (1999), Cluster Analysis and Forest Industry Complex, Research Institute of the Finnish economy, Discussion Paper no. 682, 1999.

Ernst&Young (2007), At the Crossroads. Global Pulp and Paper Report 2007 (Helsinki: Ernst&Young).

Fil'tsov, Roman and Milyaev, Pavel (2005), 'Ost-Vest Grup" otstoyala Volgu', *Vedomosti*, 88 (1369), May 18.

Fëdorov, Ivan (2003), 'Prodazha kak sposob zashchity', *Lesnaya novosti*, 16, September 25.

Grishkovets, Yekatarina (2007), 'Amerikantsy dovedut "Ilim" do gotovoy produktsii', *Kommersant*, 148, August 20.

Ilim Pulp Enterprise <http://www.ilimpulp.ru>.

Investlesprom <http://www.investlesprom.ru>.

Jensen, Michael C. (1986), 'Takeover controversy: analysis and evidence', *Midland Corporate Financial Journal*, vol. 4, No. 2, Summer 1986. Available at SSRN: <http://ssrn.com/ abstract=173452>.

Jensen, Michael C. (1988), 'Takeover their causes and consequences', *Journal of Economic Perspectives*, Winter 1988, vol. 2, No. 1, pp. 21–48. Available on the Social Science Research Network (SSRN) Electronic Library at: <http:// papers.ssrn.com/ ABSTRACT=173455>.

Kistanov, Mikhail (2006), 'Bumazhnaya voyna', *Izvestiya* (Rossiya), 72 (27113), April 24.

Khakimov, Timur (2004), 'Vrazhdebnye pogloshcheniya: Tekhnologiya, strategiya i taktika napadeniy', *Imushchestvennye otnosheniya Rossiyskoiy Federatsii*, April 26.

Khrennikov, Il'ya (2003) 'Lesnye voyny prodolzhatsya', *Vedomosti*, 237 (1037), December 25.

Khrennikov, Il'ya, Shchrbakova Anna and Sikamova, Anzhela (2004), 'Novyy vitok "lesnykh voyn" Mozhet nachat'sya posle pokupki strukturam "Bazela" 12,5% aktsiy Arkhangel'skogo CBK', *Vedomosti*, 37 (1077), March 4.

Khrennikov, Il'ya and Simakov, Dimitriy (2005), 'Novaya Mishen' "Bazela" kholding Olega Deripaski zainteresovalsya CBK Volga', *Vedomosti*, 12 (1294), January 26.

Klimenko, Jaroslav (2003), 'Voyna bez pobediteley', *Ekspert*, 48 (403), December 22.

Klyuchko, Vladimir (2003) 'Perekhvat upravleniya v aktsionernom obshchestve: napadenie i zashchita', *Rynok tsennykh bumag* (Moskva), 1, January 13.

Kornai, J., Rothstein, B. and Rose-Ackerman, S. (eds) *Creating Social Trust in Post-Soviet Transition*. (Palgrave Macmillan).

Kortelainen, Jarmo and Kotilainen, Juha (2003), 'Ownership Changes and Transformation of the Russian Pulp and Paper industry', *Eurasian Geography and Economics*, 2003 44, No. 5, 384–402.

Legrand, Pierre (1996), 'European Legal Systems Are Not Converging', *The International and Comparative Law Quarterly*, vol. 45, No. 1. (January, 1996), 52–81.

Lesgazeta.ru (2007) 'Bank Moskvy formiruet kholding', *Rossiyskaya lesnaya gazeta, Lesgazeta.ru*, 5–6 (187–188), February 11.

Loginov, Alekseiy and Tormozova, Tat'yana (2007), 'Korporativnye konflikty i reyderstvo v Rossii', *Rynok tsennykh bumag* (Moskva), 12, June 30.

Markov, Pavel (2007), 'Sovokupnost' problem teorii i praktiki nedruzhestvennykh pogloshcheniy', *Pravo i Ekonomika*, 2, February 28.

North, Douglass, C. (1990), *Institutions, Institutional Change and Economic Performance*. (Cambridge: Cambridge University Press).

North, Douglass, C. (1997), 'The contribution of the New Institutional Economics to an Understanding of the Transition Problem', WIDER Annual Lectures March 1.

Novolodskaya, Svetlana and Temkin, Anatoliy (2002), 'Ilim Palp teryaet CBK i ozhidaet silovogo zahvata kombinata', *Vedomosti*, 117 (680), July 9.

Pozdnyakov, Aleksey (2006), 'Mozhet li zahkvat byt' druzhestvennym?', *Argumenty i fakty*, 40, October 4.

Pravda Severa (2004), 'V khode shturma Sokol'skogo CBK postradali devyat' chelovek', *Pravda Severa*, November 10.

Pravda Severa (2006), 'Korporativnye konflikty', *Pravda Severa*, February 8.

Pravda Severa (2007), 'Istoriya arestovannykh aktsiy ACBK', *Pravda Severa*, March 14.

Pronin, Vladimir (2003), '600 dney lesnoy voyny Bazovyy element protiv Ilim Palp enterprayz', *Slijaniya i Pogloshcheniya*, 5/6.

Protskuryakova, Juliya (2006), 'Reyderam zakon propisan', *Rossiyskaya biznes-gazeta*, 46 (583), December 5.

Radygin, A., Entov, R. and Shmeleva, N. (2003), 'Problems of Mergers and Takeovers in the Russian Corporate Sector', *Problems of Economic Transition*, vol. 46, no. 7, November 5–64.

Radygin, A.D., Gontmakher, A.E., Kyzuk, M.G., Mezheraups, I.V., Swain, H., Simachiov, Yu.V., Shmeleva, N.A. and Entov, R.M. (2005), *The institution of bankruptcy: development, problems, areas of reforming*. Consortium for Economic Policy Research and Advice, Moscow.

Rebov, Dimitriy (2005), 'Bor'ba za ACBK razgare', *Pravda Severa*, December 16.

Regnum (2005), 'CBK "Volga" prosit polpreda i FSB opedelit' rol' nizhegorodskikh vlastey v korporativnoy voyne vokrug CBK', *Regnum* 12:38 31.03.2005, <http://www.regnum.ru/ news/430585.html>.

Romanenko, Dimitriy (2007), 'Korporativnye konflikty: Mozhno li stavit' tochku?', *Rynok tsennykh bumag* (Moskva), 12, June 30.

Rossiyskaya Gazeta (2002), 'Prekratit' "lesnye voyny"!', *Rossiyskaya Gazeta*, 90, May 23.

Rossiyskaya Gazeta (2006), 'Na trope lesnoy voyny', *Rossiyskaya Gazeta*, 86 (4052) April 25.

Schwert, William, G. (2000), 'Hostility in takeovers: in the eyes of beholder?', *Journal of Finance*, vol. 55, No. 6, 2599–2640.

Sikamova, Anzhela (2004a), 'Konets lesnoy voyny. "Ilim Palp" podpisal mirovuyu c "Bazelom" i PSB (SPB)', *Vedomosti*, 201 (1241), November 2.

Sikamova, Anzhela (2004b), 'Bitva za CBK. V khode shturma Sokol'skogo CBK postradalo devyat' chelovek', *Vedomosti*, 198 (1238), October 28.

Sikamova, Anzhela (2004c), 'Novaya lesnaya voyna vozmozhno pazgopaetsya za Arhangel'skiy CBK', *Vedomosti*, 215 (1255), November 23.

Sikamova, Anzhela and Shcherbakova, Anna (2004), 'Lesnye voyny. Episod III. Osporena konsolidatsiya aktsiy Arhangel'skogo CBK', *Vedomosti*, 81 (1121), May 17.

Solombala pulp and paper mill <http://www.sppm.ru>.

Sterkin, Filipp (2007), 'Zakon reyda,' *Vedomosti*, 45 (1819), March 15.

Temkin, Anatoliy (2004), 'Lesnye voyny ne kontshayutsya', *Vedomosti*, 50 (1090), March 25.

Temkin, Anatoliy (2005a), 'Novaya lesnaya voyna Bazela', *Vedomosti*, 161 (1442), August 31.

Temkin, Anatoliy (2005b), 'Novaya ataka na ACBK. U Pulp Mill arestovany 16% aktsiy kombinata', *Vedomosti*, 226 (1507), December 1.

Temkin, Anatoliy (2006), 'Aktsii ACBK ostalis' pod arestom', *Vedomosti*, 30 (1157) February 21.

Teubner, Gunther (1998), 'Legal Irritants: Good Faith in British Law or How Unifying Law Ends up in New Divergences', *The Modern Law Review*, vol. 61, No. 1. (January, 1998), 11–32.

Tompson, William (2003), 'Reforming Russian Bankruptcy Law', *International Company and Commercial Law Review 4*, Sweet & Maxwell. <http://www.bbk.ac.uk/polsoc/staff/academic/bill-tompson/reforming-russian-bankruptcy-law>

Volkov, Aleksandr, Gurova,Tat'yana and Timov, Viktor (1999), 'Sanitary i marodery', *Ekspert*, 8 (171), March 1.

Volkov, Vadim (2004), 'The Selective Use of State Capability in Russia's Economy: Property Disputes and Enterprise Takeovers, 1998–2002', in Kornai et al. (eds).

Watson, Alan (1993), *Legal transplants. An approach to comparative law*, 2nd edition. (Athens: University of Georgia Press).

Yeremenko, Valeriy (2004), 'Istoriya odnogo konflikta', *Upravlenie kompaniey*, 1 January 26.

PART II
Case Studies on Different Aspects of Governance

Chapter 7

Re-Territorializing the Russian North Through Hybrid Forest Management

Juha Kotilainen, Antonina A. Kulyasova, Ivan P. Kulyasov
and Svetlana S. Pchelkina

Introduction

The Russian North is currently a site of not only extensive natural resource extraction but also conservation. During the past decade, the region has become a contested ground for actors from various parts of the globe who are involved in activities related to natural resources and the environment. These include governmental actors, business enterprises and non-governmental organizations. As a consequence, the Russian North has in many ways become fundamentally linked to the rest of the world through sociospatial processes and actors' involvement in natural resource extraction and conservation. The paper argues that this involvement has resulted in a process of re-territorialization.

By focusing on the environmental and sustainability politics related to the forests of the Russian taiga, the paper seeks to examine how transnational processes modify local socioeconomic and ecological situations and how these transnational processes themselves change in the new circumstances. We wish to shed light on the issue of whether the Russian North is being divided into new territorial formations according to new environmental and social standards that are being imported into Russia and, in the process, mixed with norms, discourses and practices at the local, regional and federal scales. While the social processes affecting and modifying the relations between environmental politics and forest governance are not territorial, these processes produce new territorial formations that constitute a fragmented landscape of environmental reform. Yet, the resulting reformulation of forest governance and environmental politics treats people and their environments differently, as there are many places that are left on the fringes of the newly shaped environmental political networks.

Our overall aim is to investigate the changing forest governance in Russia. On a more concrete level, a crucial element in the transformation of forest governance is the transformation of forest management regimes. By 'forest management regimes', we understand the societal system that exists for the purpose of maximizing the benefits from the utilization of forests (cf. Eikeland and Riabova 2002). What these maximal benefits are is a debatable issue not only within but also outside the established governance systems. Thus, a particular management system is potentially in a state of constant transformation, although normal resistance to change is naturally present, as in any system. We wish to argue that while the former forest management regime, that deriving from the Soviet period, has been eroded, what we see in the early twenty-first century is a fragmented landscape of forests, forestries and local socioeconomies.

In order to understand exactly how the new fragmented territorial landscapes are being produced, we have to explore the socioeconomic processes causing the fragmentation. Our argument in this paper is that the new forest management regimes that are being produced through changes in the larger forest governance system have a hybrid character.[1] We understand this hybridity in three senses. Firstly, by 'hybrid management regimes' we mean the new, strong involvement of other than state actors – business enterprises and non-governmental organizations – in the regulation of forest management and use. Secondly, we understand the term to mean the mixing of norms, discourses and practices deriving from the Soviet period with those brought to Russia with new market economic relations, capital flows, and non-governmental organizations (cf. Kosonen 2005). Thirdly, we conceive of hybridity as a multi-scalar characteristic of forest management regimes.

A more specific issue within forest governance and management is forest certification. As discourse and practice, forest certification is a phenomenon that has been rapidly expanding across the globe (see e.g. Bartley 2003). In short, forest certification creates and implements a specific system for controlling the quality of the ecological, social and economic aspects of forest utilization. The appearance of forest certification can be attributed to the influence of West European markets in particular (specifically, those in countries of the European Union, such as Germany, the Netherlands and the United Kingdom) on the quality of the raw material used by the forestry industry to produce goods used in those countries. Moreover, new public policies are being created in some countries of the European Union according to which products purchased by public authorities must meet certain certification criteria. Different forest certification systems have been initiated by a variety of actors representing non-governmental organizations, business enterprises and governments. Consequently, different – and at least partly competing – systems for the certification of forests exist, and their implementation varies greatly from country to country. Even so, a certain degree of sustainability is required as a common denominator from the production process within these

---

1   Hybridity is a concept that has become very popular recently (see e.g. Whatmore 2002). One of its main sources of popularity was Latour's (1993) book on the peculiarities of the 'modern constitution', in which he maintained that modern science and thinking is causing us to see things as bi-polar constellations, one of the poles being nature, the other culture. Latour stated that the world in fact consists of all sorts of 'hybrids' that are becoming more and more difficult to explain by modern science, which derives its essence from the scientific revolution of the eighteenth and nineteenth centuries. In the present chapter, our perspective on 'hybrids' is somewhat more limited.

systems. The Forest Stewardship Council (FSC) system for forest certification is by far the most prominent of the systems in use (see e.g. Kotilainen et al. 2008). As there are products on the market made from raw materials originating in Russia, the requirements and impacts of certification extend to Russian territory.

As a specific feature within changing forest governance and management, we therefore focus on the processes of implementation of certification in contemporary local settings in the Russian North. In particular, we wish to shed light on the issue of how a supra-local socioeconomic and ecological process such as forest certification is being adopted, interpreted and developed in the regional and local contexts in Russia. We are particularly interested in how Russian enterprises, embedded in regional and local contexts, function under transnational influence. Moreover, we wish to explore how the changing forms of forest management – and, more specifically, the certification processes – affect the lives and livelihoods of the residents of local forestry settlements. Furthermore, we analyze the factors that produce differences in terms of local outcomes between similar cases. In other words, by closely examining nodes in a changing network of forest governance, we wish to provide a detailed picture of how the new governance mechanisms operate in relation to older mechanisms deriving from previous periods and governance styles. As a reaction to the new requirements, several different systems for forest certification have already been proposed for Russia (Tysiachniouk 2006). However, the Forest Stewardship Council (FSC) certification system is currently the most widely used in the country, albeit on limited forest territories. For example, Ilim Pulp Enterprise, a major player in the Russian forest industry, has recently decided to adopt the FSC system. The Russian national certification system is currently being tested, as is illustrated by one of our cases presented below.

While the rapid restructuring of the forest management regimes in Russia in recent years has included changes in the state structures for forest governance, transformation of the forest industry as a sector of the economy and the introduction of new external influences in Russia, these transformations have had spatially varying consequences in the turbulent post-Soviet conditions. Accordingly, it is our aim to analyze the implications that these transformations have had on a local scale. Specifically, our focus is on the diverging development of logging enterprises, *lespromkhozy*, and the settlements in which these enterprises are located. In order to demonstrate changes in forest management practices and discourses and to illustrate how forms of hybrid management are produced, we will examine four cases from the Russian North. While the case enterprises and localities differ in terms of volume of production, number of workers, and number of inhabitants, the enterprises are all mainly engaged in timber cutting, have similar conditions of management, belong to larger holding companies and play significant roles in the local and regional (*subject of the federation*) economies. Three of the enterprises possess the international FSC certificate; one has received the Russian national certificate, 'the national system of voluntary forest certification in Russia', which was developed in 2001 by the National Council for Voluntary Forest Certification in Russia in a process headed by Academician G.A. Rakhmanin. The creation of the system was initiated by M.V.

Tatsyun and G.A. Gukasyan, who represent the Russian Forest Industrialists and the Forest Experts Union, respectively, and the work was carried out with the support of the Forest Industry Department of the Ministry of Industrial Sciences of the Russian Federation. In 2004, an experimental audit was conducted at the joint-stock company Vozhegales. While very similar to the FSC system, the Russian national system does not include the controlling instruments of the FSC, and it has not yet been legitimized on the internal and external markets.

By using these locations as focal points within the changing forest management regimes, we wish to increase the understanding of the processes through which non-local initiatives intermingle with local conditions and in effect produce hybrid forms of ideas and practices in forest utilization. Accordingly, the chapter analyzes the 'migration' and introduction of external environmental and social requirements into the post-Soviet context in Russia. We will firstly explore the longer historical and geographical characteristics of the enterprises and localities; secondly, explore the consequences of post-Soviet economic transformations on the local scale; thirdly, scrutinize the hybrid forms of economic practices there; and, finally, investigate in more detail the effects of forest certification for the localities. We will conclude by arguing that the processes underway are producing a new territoriality, creating diverging and fragmented forest, social and economic landscapes within the Russian Federation.

Local Historical and Geographical Characteristics

The Russian North has a lengthy history of forest use. This history can be traced back to the Middle Ages, but what is important for an understanding of the current situation is the Soviet history with its large-scale industrialization projects that included forced movement of people across the Soviet space (see Moran 2004). The Arkhangel'sk and Vologda Regions in the north of Russia are forested regions where timber cutting and processing are crucial to the local economies and important even for the regional economies as well. The joint-stock company Maloshuykales is located in the settlement of Maloshuyka (Onezhskiy district, Arkhangel'sk Region), the Dvinskoy forestry enterprise in Dvinskoy (Kholmogorskiy district, Arkhangel'sk Region), the Belozerskiy forestry enterprise in the town of Belozersk (Vologda Region), and the joint-stock company Vozhegales in the settlement of Kadnikovskiy (Vologda Region). The Dvinskoy and Maloshuyka settlements were constructed during the Soviet era, whereas Kadnikovskiy dates back to the early twentieth century and was constructed during the expansion of the timber industry into previously uninhabited places. The decisive arguments for choosing these places for settlements were their proximity to the forests to be cut as well as the availability of transportation facilities. The town of Belozersk has a history going back more than 1,000 years and while forest exploitation has always played an important role in its development, the forest industry was, obviously, not the initial reason for its creation.

Table 7.1   Main Data Concerning the Localities Studied

Settlements/data	Dvinskoy	Maloshuyka	Belozersk	Kadnikovskiy
Time of creation	Soviet period, 1950s	Soviet period, 1950s	has existed more than 1000 years	Before the 1917 Revolution
Means of transportation	Severnaya Dvina river, White Sea ports, main road	Railway	Main road; Lake Beloye (connected to the Volga-Baltic waterway system)	Railway
Number of inhabitants	1,000	2,000	12,000	2,000
Origin of population	Nearby villages	Various republics of the former USSR	Local people and those from the nearby villages	Vologda Region, various republics of the former USSR

In Maloshuyka and Kadnikovskiy, the railway provides the main means of transportation, as automotive transport is difficult – or even impossible – during autumn and spring. In winter, marshy lands and waterways freeze, becoming available for motor transportation. In contrast, the riverside settlement of Dvinskoy has no railway connection. The district centre, Belozersk, is a small town with well-developed motor transport communications as well as a port on Lake Beloye, which connects the town to the Volga–Baltic waterway. The absence of roads, with the railway being the only means of timber transportation, has, during hard times, complicated the selling of products from Maloshuyka and Kadnikovkiy, as the enterprises were dependent on railway tariffs and on the conditions for and admissible volumes of goods for transportation. On the other hand, the presence of the railway junction as a second source of jobs in Maloshuyka helped the local community survive the period of crisis in the 1990s. In Dvinskoy and Kadnikovskiy, the only source of work was lespromkhozi, and the crisis affected the local population more adversely. In the district centre, Belozersk, there were many jobs financed from the budget of the company administration and, furthermore, the Belozersk forestry enterprise managed to operate in the 1990s without heavy losses.

The inhabitants of Maloshuyka originate from various republics of the former USSR and their descendants, while the Dvinskoy settlement is mostly populated by inhabitants of the neighbouring villages. In the settlement of Kadnikovskiy, the inhabitants originate from the nearby villages and other parts of the former Soviet Union. In Belozersk, local residents and people who have moved from nearby villages make up the majority of the population. These variations in the origins of the populations partly explain the differences in their attitudes towards the post-Soviet transformations. Maloshuyka's inhabitants have shown great social apathy because they still feel they are 'not at home' as people who 'came in search of a job' and were 'forced' by the social and economic crisis to stay. The residents of the other settlements do not have such an attitude, as they are more original inhabitants or come mostly from the nearby villages.

In the Soviet times, most of the social infrastructure of the three rural settlements and a considerable part of the infrastructure in the town of Belozersk were financed and supported by the lespromkhozy. At that time the Belozersk forestry enterprise, Maloshuyka forestry enterprise (later re-organized as Maloshuykales), and Mitinskiy forestry enterprise (reformed as Vozhegales) were large and economically thriving companies. The Dvinskoy forestry enterprise was rather small by comparison.

Local Effects of the Reforms of the 1990s

The transformation of the Russian economy in the early 1990s created economic problems in the focal lespromkhozy. Supporting the infrastructure became a heavy burden for the enterprises, and the Maloshuyka, Dvinskoy, and Mitinskiy forestry enterprises all incurred debts. The Belozersk forestry enterprise also experienced difficulties, although not to the same extent as the others.

Although all Soviet lespromkhozy were reorganized into joint-stock companies in the early 1990s, the ways in which they developed thereafter varied, as will be illustrated below. The Maloshuiskiy forestry enterprise maintained its ties with the Onezhskiy forestry enterprise, to which it used to deliver round timber, but because of its large debts it could not function well, update its equipment or pay salaries, and ultimately went bankrupt. In 1997 a new enterprise, an open joint-stock company, Maloshuykales, was established. Its shareholders bought the majority of the stock and the leasehold of the Maloshuiskiy forestry enterprise. The Onezhskiy forestry enterprise became its main shareholder. After this, the Maloshuyka forestry enterprise operated for a few more years, went bankrupt and was closed. Thus, a market economy enterprise took the Soviet enterprise's place, and its major shareholder interested in raw material deliveries got the new viable enterprise free of debt and formal responsibilities where the maintenance of the infrastructure in the settlement of Maloshuyka was concerned.

The Mitinskiy forestry enterprise, which also was a joint-stock company in the early 1990s, had the same problems of not being able to pay salaries and taxes or maintain the local social infrastructure and experiencing a drop in profitability. It could not survive in its old form either. In contrast to the Maloshuiskiy forestry enterprise, its business ties had been cut. In 1997, in order to survive in the new circumstances created by the market economy, the administration of the Mitinskiy forestry enterprise made a decision to divide it into two new companies. The larger of these, Vozhegales, in practice kept the workers and directors of the former enterprise and started confronting the crisis without any external support. Only in 1999 did it choose to join a larger holding company, since it believed that this was the only way to operate profitably. These two cases show that at this stage of reformation quite large lespromkhozy – in terms of volume of cutting, number of workers, accompanying industries and the social infrastructures of the associated forestry settlements – either went bankrupt or were transformed through fragmentation into newly created enterprises.

The business ties of the Dvinskoy forestry enterprise were also cut. Like Vozhegales and most forestry enterprises, Dvinskoy could not survive without the help of a large shareholder interested in a stable supplier and able to invest in its production. The enterprise Dammers, founded in Arkhangel'sk by the German company Holz Dammers Moers, became such a shareholder. The Dvinskoy forestry enterprise did not go through the bankruptcy procedure and therefore still has some debts and has partly continued to maintain the social infrastructure in the Dvinskoy settlement.

In contrast to the three previous cases, the Belozerskiy forestry enterprise, the largest of the four companies, neither went bankrupt nor became part of another company. The relatively stable operation of this enterprise during the reform resulted from a prescient economic policy and management system developed by the forestry industrial association Cherepovetsles. The Belozerskiy forestry enterprise had been a part of Cherepovetsles even during the Soviet era. The association combined timber cutting and woodworking enterprises in several districts of the Vologda Region. After the beginning of the reforms, all logging enterprises, including the Belozerskiy forestry enterprise, were reorganized into joint-stock companies and remained members of Cherepovetsles. Currently, Cherepovetsles is a forestry holding company. Although large timber industry enterprises dropped out of it, Cherepovetsles managed to arrange profitable sales, also to Western markets. It was, for example, able to construct two sawmills in the early 2000s.

The differences in the reorganization of the enterprises at the beginning of the century can partly explain the differences in their later development. Thus, Maloshuykales has been paying salaries and taxes regularly, and its principal shareholder has been updating the equipment and investing in modern timber cutting equipment, such as harvesters and forwarders. In contrast, until 2005, the Dvinskoy forestry enterprise had certain problems with timely payment of salaries. The principal shareholder of the enterprise was not investing in new equipment and it has often had problems with timber cutting. Vozhegales, as well, has not been receiving direct investments from its holding company, but it has been renewing its machinery as much as possible at its own expense, admittedly using old-design Russian equipment. Cherepovetsles, which has a common management system for the whole company, contributed to the regular saving of resources at the Belozersk forestry enterprise, which allowed for investments in new equipment and technologies. It is evident that the social and economic situation of the Belozersk forestry enterprise is the most stable, with Maloshuykales, Vozhegales, and Dvinskoy following in this order. Another reason for the stability of the Belozerskiy forestry enterprise, Maloshuykales and Vozhegales is their long-term forest lease of 49 years. This is also a key condition for FSC certification. When the Dvinskoy forestry enterprise applied for certification in 2000, it had a lease

Table 7.2   Data on the Enterprises Studied

Enterprise/ characteristics	Dvinskoy	Maloshuykales	Belozersk forestry enterprise	Vozhegales
Soviet enterprises first organized as	Dvinskoy forestry enterprise, 1955	Maloshuyka forestry enterprise, 1950	Belozersk forestry enterprise, 1929	Mitinskiy forestry enterprise, 1930s
Reformed enterprise created	1992	1997	1992	1997
Principal shareholder	LLS Dammers, since 1995	Joint-stock company 'Onezhskiy forestry enterprise', since 1997	Joint-stock company forestry corporation 'Cherepovetsles' since 1992	Holding company 'Vologda forest industrials' since 1999
Production ties with the principal shareholder before privatization	None	Traditional, Onezhskiy forestry enterprise had connections with Maloshuyka forestry enterprise	Traditional, with Cherepovetsles since Soviet times	None
Shareholder	Holz Dammers Moers (Germany)	Russian trust 'Orimi'	None	None
Woodworking	Log sectioning plant	Sawing for local purposes, does not cover the needs of the settlement	Large log sectioning plant	Sawing for local purposes, does not cover the needs of the settlement
Annual volume of timber cutting	70,000–100,000 m^3	140,000 m^3	600,000 m^3	370,000–400,000 m^3
No. workers 2003	140	350	1,570	710
Main buyer of round timber	LLS 'Dammers'	Joint-stock company 'Onezhskiy forestry enterprise'	Metsäliitto and UPM Kymmene (Finland), Stora Enso (Sweden-Finland), Nivida (Sweden), Telemark Wood Com. As. and Södra Skog (Norway)	Sales through trading company JSC 'Astrofor', which belongs to holding company 'Vologda Forest Industrials'
Main buyers of round timber and timber through the holding companies	Holz Dammers Moers (Germany) local enterprises in Arkhangel'sk	Western Europe	Only sawing log Molfenter & Co (Germany), Satim (Netherlands), Barthel Pauls and Sinbpla (Belgium)	Thomesto/ Metsä-liitto (Finland), LLS 'Kharovsklesprom'
% taxes paid by enterprise in district budget	Insignificant	Insignificant	46%	70%

of five years; this expired in 2003 and was continued in 2004 for five years, but its further extension is an open question. As a result, the Dammers enterprise, as a shareholder, considers investing in new equipment at the Dvinskoy forestry enterprise quite risky, as it might lose its leased forest areas at any moment and go bankrupt.

Hybrid Economic Practices Mixing Soviet and Post-Soviet Systems

It can be argued that as a result of the restructuring and transformation processes described above, new blends of economic practices that mix Soviet and post-Soviet forms of operation have emerged at the local scale in the Russian North. This hybridization is seen, first, in the rigid vertical structure of the new holding companies, in which the new governing hierarchies are in certain ways similar to the Soviet production associations. While the Maloshuykales, Vozhegales, Belozersk and Dvinskoy forestry enterprises are formally independent, they are, in fact, sub-divisions of larger companies that own a controlling majority of the shares. As a rule, forestry enterprises belonging to a forestry holding company are organized in legal terms as joint-stock companies. However, their economic, social and environmental policies are defined by their owners. The management structure is designed in such a manner that all economic activities, especially wages and investments, are under the monthly control of their main shareholders. At the end of every month, all bookkeeping records of the enterprises are transmitted to the top-level management for approval.

Cherepovetsles, the owner of the Belozersk forestry enterprise, has a system of corporate standards. Non-compliance with these regulations is unfavourable for the local forestry enterprise, as it adversely affects the wages of employees, including those of the managing staff, and affects other issues at the enterprise as well. As a general rule, members of the board of directors of the holding company are also members of the board of directors of the local enterprise. For example, the acting general director of Cherepovetsles is at the same time the president of the board of directors at the Belozerskiy forestry enterprise. In the case of the Dvinskoy forestry enterprise, the review is even more rigorous: all aspects of management are supervised, including human resource policy, everyday economic activities and work discipline.

At the same time, there is variation in the situation at the four enterprises. As distinct from the Dvinskoy forestry enterprise, the management of the Belozersk forestry enterprise, Vozhegales and Maloshuykales is stable. At Belozersk and Vozhegales, the directors have not changed from the Soviet times. In the case of the Belozersk forestry enterprise, the director of the former Soviet Belozersk lespromkhoz heads the new enterprise. At Vozhegales, one of the directors of the Mitinskiy forestry enterprise became the director of the new company. The director of Maloshuykales also used to work as a forest expert at the Maloshuyka forestry enterprise. All of them, as local residents, have sought to keep the enterprise's workforce and the community associated with the enterprise, which was loyal to them, proud of the history of the company and not to divide it into Soviet and post-Soviet periods. They consider that even though the enterprise goes through different phases, it nevertheless maintains its traditions. Moreover, these directors enjoy the confidence of the local residents and do not separate themselves from them. Furthermore, they are quite independent of the holding companies in deciding on forest management issues. In contrast, at the Dvinskoy forestry enterprise, ten directors appointed by Dammers have been changed during the last ten years. The majority of these directors have not had any special forestry background or been local residents. They have not been able to make decisions without approval from the Dammers Board of Directors, and they have not had authority over the workers in the same way as the directors in the other enterprises. Therefore, the Dammers management at the Dvinskoy forestry enterprise has often been rather inefficient. Overall, these cases show that there are administrative and executive mechanisms in joint-stock companies that to some extent correspond to the hierarchical subordination in enterprises in Soviet times.

Another practice that can be seen as a hybrid form of Soviet and post-Soviet governance is the paternalism exercised by enterprises in local contexts. As mentioned above, in Soviet times the social infrastructure in the forestry settlements was owned by the forestry enterprises and was financially supported by them. This local paternalism corresponded to the generally paternalist structure of the Soviet state. In Soviet localities, this structure was built up by both management at enterprises, which bore all responsibility for the functioning of the settlement, and the inhabitants, who considered the enterprise as their only source of livelihood. During the post-Soviet transition, these paternalistic relationships between enterprises and forestry settlements have not completely disappeared, although they have been undermined.

For example, during interviews, respondents from Maloshuyka and Kadnikovskiy recalled that in Soviet times wages were high, workers were provided with what were known as deficit goods,[2] and workers often received sanatorium and spa treatment guaranteed by the local trade union. Respondents recollect those years with nostalgia; they believe that things were formerly 'very good'. Consequently, they think that the enterprise should still support their lives; their requirements concerning the social programmes provided by their enterprises are thus very high, since they compare them with the programmes of the Soviet times. Slightly in contrast, the inhabitants of Dvinskoy treat the Soviet period as the time when they were always provided with jobs and stable salaries, although they do not have illusions about a 'lost paradise'. Hence, their demands of the enterprise are clearly formulated and focus on stable employment and timely payment of salaries. In all the localities, however, the local residents make social demands of the respective enterprises. The management of the enterprises, on the other hand, also continues to reproduce elements of the Soviet-era relationships. Furthermore, companies such as the Belozersk forestry enterprise, which provides about half, and Vozhegales, which provides about 70 per cent, of the budget of

---

2   These were imported goods that could not be bought in shops; they were purchased by the Ministry for Forest Industry and distributed amongst the workers in the industry.

their respective districts, continue to support the economy of the district and the residents' welfare. This strengthens the reproduction of the paternalistic socioeconomic construction.

Surprisingly enough, the social requirements of the new forest certification systems – the FSC and the Russian national system of Voluntary Forest Certification – often coincide with Soviet conceptions of responsibility of the enterprise for the welfare of local communities. The directors of Maloshuykales, Vozhegales and the Belozersk forestry enterprise consider themselves to be local residents and thus part of the local community. They have kept their 'Soviet ideas' and believe that in the North the enterprises must, for example, provide the population with firewood either for free or for very low prices. Such ideas also correspond to the social requirements of the FSC. Consequently, a long-term social programme – 49 years, corresponding to the duration of forest lease – has been developed at Maloshuykales. At the Belozersk forestry enterprise a similar programme is now under development. In this case, the requirements of the FSC and the interests of the holding company in the certification of its enterprises played major roles in the decision to adopt the long-term programme. For example, at Maloshuyka there is a plan for social development until 2052 that is coordinated with the local public administration. According to the plan, the enterprise supports the professional education of new workers and recruits only from the local population. This decision has been confirmed by refusing to hire teams of loggers from other regions of Russia. In this case, we notice that the social requirements of the FSC voluntary forest certification, the ideas of the management at Maloshuykales and the needs of the local community have coincided. At Vozhegales, in spite of the long-term lease, such a programme has not been developed and adopted because, firstly, it was not required by the Russian national certification system and, secondly, the holding company's interest in certification was low. Moreover, in accordance with the requirements of the FSC, Maloshuykales and the Belozersk forestry enterprise have made efforts to develop the local population's social responsibilities. At Maloshuykales, the creation of a trade union was initiated, a forestry and environmental club was established for children at the local school and several educational events for workers were organized. The Belozersk forestry enterprise and Vozhegales have also organized educational events and supported environmental initiatives and trade unions.

At the Dvinskoy forestry enterprise many directors have been replaced and all of them have been quite formal figures in contrast to the other settlements. Therefore, we can only refer to the conceptions of a former general director of Dammers in Arkhangel'sk, who in 2005 became the general director of the Dvinskoy forestry enterprise. In fact, he has been the leader of the Dvinskoy enterprise in practice since the late 1990s. Living in Arkhangel'sk, he does not consider himself to be a local resident in Dvinskoy. He has stated that the weaknesses of local authorities and their dependence on the sole enterprise in the surroundings have forced him to support the social structures of the settlement and, accordingly, many social responsibilities have remained concerns of the enterprise. However, his conception of social support does not always coincide with the residents' expectations. The Dvinskoy forestry enterprise supports key social infrastructure, including the ferry, fire department, boiler-house, bakery, health care centre, shop, school, clubhouse and library. Such social obligations are rather onerous for the owner: only the bakery and the shop bring in income; the other units operate on subsidies. The director regularly raises the prospect of terminating their financing, offering instead to help in the management and transformation of these unprofitable activities into profitable ones. However, as this is bound to lead to an increase in the prices of services, he has not received support from the local residents.

Actors' Roles and Interactions in Introducing Transnational Socioenvironmental Imperatives into the Local Contexts

Finally, we explore the roles of different actors in introducing forest certification into the Russian North. In the cases studied here, forest certification was initiated by the owners of the forestry enterprises, that is, the holding companies Cherepovetsles, Vologda Forest Industrialists, the German company Holz Moers Dammers and the Russian concern Orimi. Their motivation was to stabilize their exports to the Western market, which has been considered environmentally sensitive. In the case of Vozhegales, certification was an experiment, as the national system of voluntary forest certification in Russia had not yet been officially registered in the country or accredited internationally. Initially, the idea of certification at Vozhegales was proposed by the National Council for Voluntary Forest Certification in Russia and the Association of Russia's Forest Industrialists and Forest Exporters. Being a member of this association, the holding company Vologda Forest Industrialists recommended that Vozhegales undergo national certification, but the company itself did not take part in the process.

In the case of the Dvinskoy forestry enterprise, the head company, Holz Dammers Moers, appointed the company that was to conduct the certification audit. At Maloshuykales and the Belozersk forestry enterprise, the designation of the auditing company and the preparation for certification were taken care of by the Onezhskiy LDK and Cherepovetsles, respectively. They cooperated with the Arkhangel'sk Centre for Voluntary Forest Certification in Russia, which was created at the Northern Scientific Research Institute for Forest Economy. At the Belozersk forestry enterprise and Maloshuykales the preparation for certification was more detailed, with local features being taking into account. Maloshuykales was finally certified in 2003, and the Belozersk forestry enterprise in 2004. Dvinskoy was certified in 2000, but with many critical observations. At Vozhegales there was no preparation for the audit. This enterprise was selected for the experiment because it was considered to be one of the best-functioning forestry enterprises in the Vologda Region. Its economic and social indices were suitable for the experiment, and the control executed by its Finnish trading partners contributed to its observance of Russian forest legislation and environmental standards, the legality of the origins of timber, and the protection of workers. The certificate was issued in 2004 with a few observations.

Through these examples, we can see the interaction of different actors. For example, the Belozersk and Dvinskoy forestry enterprises and Maloshuykales carry out selective felling, which the system of voluntary forest certification in Russia encourages. This is in stark contradiction with the viewpoint of the regional forestry authorities, who often do not give permission for selective felling but instead require that clear-cutting be carried out. These authorities refer to Soviet scientific research and post-Soviet norms and standards based upon them. Clearly, such a policy pursues short-term objectives, such as an increase in taxes received from the enterprises, preservation of jobs, providing regional pulp and paper mills with raw material and producing cheap firewood for the local population. Such discrepancies between Russian harvesting norms and rules and the international standards of the FSC may even lead to conflicts between enterprises and state bodies. However, the heart of the conflict may lie in the difficulty of incorporating the enterprise into the regional network of governance rather than in contradictions between Russian and international norms and rules.

The influence of international NGOs on the practices of certified enterprises is so substantial that the enterprises sometimes even incur certain economic losses to retain their image and the opportunity to work on the Western markets that they consider to be environmentally sensitive. For example, all four companies studied here signed an agreement with Greenpeace on a five-year moratorium on felling in old-growth forests or forests with a high nature conservation value that are situated in their leased territories. Furthermore, the enterprises have planned to save parts of these forests as widened water protection zones after the moratorium ends. Nevertheless, the enterprises keep on paying rent for these sites as if they could carry out felling in them. Even if they have agreed with Greenpeace on the value of these forests, they consider them 'mature', that is, forests that can become 'overmature, and perish'. Therefore, recognizing Greenpeace's and environmental experts' international authority, they consider such forests as 'reserves' or 'sites of limited forest exploitation'. Moreover, some enterprises even finance research. For example, Maloshuykales finances research on rare flora and fauna on its leased territories in order to pick out small forest sites and save key biotopes. In 2004, it withdrew 28,000 cubic metres of timber for these purposes from its planned felling areas. This territory was transferred into the state forest fund and received the status of protected nature territory. However, the responsibility for the protection of this forest still lies with Maloshuykales.

Independent rating agencies such as Expert RA[3] have noted the process of incorporation of Russian certified enterprises into global networks of non-state governance. Onezhskiy LDK and Cherepovetsles received high environmental ratings. In the beginning, Dammers was low in the ratings because the certificate of the Dvinskoy forestry enterprise was suspended in 2002 due to its failure to pay

---

3   www.raexpert.ru.

salaries in a timely manner and its violation of the rules of forest exploitation. In 2003, the certificate was renewed, as the environmental requirements had been met and the wages in arrears had been paid. In 2005, the rating of Dammers went up because it had again passed the FSC primary audit and considerably improved its economic, environmental and social indices. The enterprise also became a member of the 'Union of Environmentally Responsible Wood-Cutters' created by WWF Russia.

Another aspect of forest certification that has impacts at the local scale is the social requirements of the certification systems. These assume public participation in decision making concerning forest exploitation, informing the local population on the social, economic and environmental aspects of certification, improvement of labour conditions and regular payment of wages. In the Dvinskoy, Maloshuyka, Belozersk and Kadnikovskiy settlements, the certified forestry enterprises strive to inform the workers and local population of the environmental requirements of certification. The Belozerskiy and Dvinskoy forestry enterprises have succeeded better than Maloshuykales and Vozhegales in this respect. The majority of workers at the Dvinskoy and Belozerskiy forestry enterprises could tell about and explain the FSC economic and environmental requirements, whereas the workers at Maloshuykales and Vozhegales were only informed of the enterprise's certification. However, the Dvinskoy forestry enterprise paid special attention to providing information only after its certificate had been suspended. This occurred because of problems with the company management and a gross violation of the rules of forest exploitation by the workers. As for the public in the settlements, the local population did not take part in working out new economic, social and environmental policies for any of the enterprises; they did not know about the social requirements of forest certification or about the procedure of public participation. Due to glaring violations of labour legislation and irregular payment of wages, the situation in Dvinskoy has changed. The local population started a dialogue with the enterprise, discussing the issues they were interested in through consultations with stakeholders.

Conclusions

As the cases have sought to illustrate, combinations of Russian and international standards of forest management have produced new hybrid forms of forest management at the local scale in the Russian North. While the practices of the market economy have been introduced into Russia's forest periphery, practices deriving from the Soviet past are still present, alive and well locally. In fact, these often do not even conflict with more recent ones. Hence, the recent restructuring processes have been hybridizing the forest management in the Russian North. First, in addition to the traditional state governance system, the introduction of enterprises and NGOs as regulators of forest use has been seen in the Russian North, albeit in limited locations and on limited territories. This is in line with the general understanding of present-day governance practices in a global sense (cf. van Kersbergen & van Waarden 2004). The traditional state governance system is being supplemented by new instruments and regulation adopted by the enterprises themselves, as well as international NGOs. In Russia, this new system functions only in terms of particular forest enterprises and in particular territorial locations, that is, where the companies are oriented towards Western markets they consider to be environmentally sensitive. However, a trend towards a supposedly rapid expansion of this phenomenon is visible.

Secondly, the Russian forest management system, partly inherited from Soviet times, has recently been accompanied by a new transnational system, introduced as a result of market influence. Accordingly, it can be concluded that hybrid forms of forest governance are being produced that are neither purely Russian nor purely Western, but constitute a new way of dealing with local socioeconomic and socioecological conditions. In the Russian system of forest governance, features of the former Soviet system are combined with those of the post-Soviet market economy, which in turn is being transformed into a new transnational system prompted by global markets and forest certification. Thus we can observe a combination of Soviet, post-Soviet and global market practices. The private market form, which supposes an independence of shareholders, nevertheless contains administrative and executive mechanisms that correspond to the hierarchic relations enterprises had with their subdivisions in Soviet times. The Soviet/post-Soviet governance hybrid practices are also manifest in the paternalism that the enterprises engage in towards the forest settlements. Thus, the hybrid governance forms clearly are not created by Russian society or Western countries alone, but represent a new way of governing the local social and environmental systems.

Thirdly, the processes that are creating the new forms of management are essentially multiscalar. Forest management and forest certification, which are part of this governance, are scaled processes. Specifically, it can be argued that the transnational FSC forest certification standards that are being localized in Russia have become a strategy of forest governance for regional forest companies and their local enterprises. As some of these standards conflict with the Russian forest legislation, the State Forest Service has sought to either find means to support this 'new useful' activity of the forest enterprises or stand up against the 'strange and illegal' practices of forest use. In both cases, the amendments have sometimes been introduced into the national or regional legislation. Such processes provide evidence of the strengthening influence of international business and non-governmental networks on the national and regional legislations. The influence of transnational processes is also seen in the creation of a Russian national system for forest certification, which is supposed to combine and accommodate the international forest certification systems, Russian legislation and the specificity of the Russian social, economic and natural circumstances. However, this system has not yet been accredited and recognized internationally; nevertheless, as a joint initiative by the Russian forest industrialists and the Russian government it is clearly a response by the Russian governance system to the global challenge posed to governance.

The new schemes of forest management are essentially shaped by new networks of forest politics (cf. Kortelainen and Kotilainen 2006). In order to be successful in forest certification, a forest company and its subsidiaries have to be involved in international and regional business networks and in international networks of environmental NGOs as members of associations of environmentally and socially responsible forest managers. As a Russian business becomes incorporated in international networks, it transfers the practices of governance created in Western countries to the regional and local scales in Russia. Nevertheless, the target audience of local businesses is not the local population and its interests, since the company introduces innovations under the pressure of environmental and social requirements of the supposed or imagined Western consumer. However, a certain fulfilment of the rights of the local residents is sometimes seen to take place, even if the population itself is not concerned with this issue.

Even though forest certification includes ecological and social dimensions, it has thus far been biased in Russia towards ecological aspects. The changes in forest governance in Russia occur mostly under the pressure of the 'Greens' and engaged investors, who are more interested in the environmental issues incorporated in forest certification than in social principles. Therefore, a company's certification may easily be suspended if environmental standards are violated, while non-compliance with social standards may pass unnoticed. Nevertheless, regional and industrial trade unions and NGOs are gradually becoming a power which will be able to put pressure on certified companies and thus also secure the fulfilment of the social standards incorporated in the FSC.

Finally, as a result of the changing forest governance, a new territoriality has been emerging in the Russian North. While the social processes affecting and modifying the relations of the spheres of environmental politics and forest governance are not territorial, new territorial formations are nevertheless produced as a result of these social processes. There are, in effect, currently different, and competing, forest management regimes in Russia. Certification areas are prime examples of the new territorialities. Consequently, a fragmented landscape of environmental reform is produced (see also Kortelainen and Kotilainen 2006; Kotilainen et al. 2008). The resulting reformulation of forest governance and environmental politics treats people and their environments differently, as there are many places that are left on the fringes of the newly shaped environmental political networks. However, it seems likely that the environmental political networks that demand explicit environmental policies from companies will continue to expand in Russia. Recently, the decision by Russia's largest forest industrial enterprise, Ilim Pulp Enterprise, to adopt the non-governmental Forest Stewardship Council (FSC) certification system is a clear sign of this process.

References

Bartley, T. (2003), 'Certifying Forests and Factories: States, Social Movements, and the Rise of Private Regulation in the Apparel and Forest Products Fields', *Politics and Society* 31:3, 433–464.

Eikeland, S. and Riabova, L. (2002), 'Transition in a Cold Climate: Management Regimes and Rural Marginalisation in Northwest Russia', *Sociologia Ruralis* 42:3, 250–266.

Kortelainen J. and Kotilainen, J. (eds) (2006), *Contested Environments and Investments in Russian Woodland Communities* (Helsinki: Kikimora Publications).

Kosonen, R. (2005), 'The Use of Regulation and Governance Theories in Research on Post-Socialism: The Adaptation of Enterprises in Vyborg', *European Planning Studies* 13:1, 5–21.

Kotilainen, J., Tysiachniouk, M., Kulyasova, A., Kulyasov, I. and Pchelkina, S. (2008), 'The Potential for Ecological Modernisation in the Russian Context: Scenarios from the Forest Industry', *Environmental Politics* 17:1, 58–77.

Latour, B. (1993), *We Have Never Been Modern* (Hemel Hempstead: Harvester Wheatsheaf).

Moran, D. (2004), 'Exile in the Soviet Forest: "Special Settlers" in Northern Perm' Oblast', *Journal of Historical Geography* 30, 395–413.

Tysiachniouk, M. (2006), 'Forest Certification in Russia', in Cashore, B., Gale, F., Meidinger, E., and Newsom, D. (eds), *'Confronting Sustainability: Forest Certification in Developing and Transitioning Countries*, Yale School of Forestry & Environmental Studies, Report Number 8 <http://environment.yale.edu/2538/confronting_sustainability_forest/>

van Kersbergen, K. and van Waarden, F. (2004), '"Governance" as a Bridge Between Disciplines: Cross-disciplinary Inspiration Regarding Shifts in Governance and Problems of Governmentality, Accountability and Legitimacy', *European Journal of Political Research* 43, 143–171.

Whatmore, S. (2002), *Hybrid Geographies. Natures, Cultures, Spaces* (London: Sage).

Chapter 8

Construction of Trust in Russian Mill Towns

Jarmo Kortelainen and Soili Nystén-Haarala

Introduction

Traditionally, single-industry towns were the only means to organize large-scale production in sparsely populated areas: it was not enough to establish a factory; a community had to be built around it and the enterprise was responsible for the community's well-being and overall viability. This was the approach taken in the Soviet Union when its resource-based industry was developed. Industrial enterprises were often set up in remote districts, with smaller communities or larger towns established around them. These communities have undergone an immense transformation in the last fifteen years, and it is these changes that we focus on in this contribution. Special emphasis is placed on communities in Northwest Russia.[1] The forest industry is the leading economic sector in the Republic of Karelia and Arkhangel'sk Region and has an important role in other parts of the region as well. In these areas, the forest industry ensures the livelihood of dozens of small and larger communities (Kortelainen and Kotilainen 2003).

In what follows, we examine how the relationship between the enterprises and local communities has taken shape in different kinds of communities. We do this by using trust as a conceptual and explanatory tool enabling us to analyze differences in company-community relations. Of the various reasons for the differentiated development of mill towns, we have chosen to investigate one factor in particular – ownership arrangements – and how this has affected the communities and their relation to enterprises. New owners represent dissimilar economic backgrounds and business cultures, which is visible locally as varying degrees of trust – and sometimes of distrust.

The contribution is based on both primary and secondary sources. The primary sources consist of semi-structured qualitative interviews of company managers and representatives of trade unions at the mills in Kondopoga and Arkhangel'sk (in June, September and November of 2004) and interviews of local and regional authorities in Kondopoga, Petrozavodsk, Arkhangel'sk and Novodvinsk[2] (in June, September and November 2004, February and November 2005). Some managers

---

1   See the map at the beginning of the book.

2   The interviews at the Arkhangel'sk paper and pulp mill in Novodvinsk are not complete. The takeover situation there, which began in 2003 and continues, has hindered interviews with managers and representatives of the municipality.

of the Segezha mill and its logging company in the village of Padani, as well as villagers who were the former owners of the logging company, were interviewed in September 2006. The primary empirical data are compared with findings of other researchers. Russian local newspapers have also provided a great deal of information about company-community relations.

Construction of Trust

Many accounts have viewed the concept of trust as an important element in explaining the construction and constancy of network relations. Our primary focus is locally specific firm-community relationships, which depend on both local and wider social relations. Thus, we will discuss both local relations and trust on a broader level in Russian society. There are approaches that focus on general trust in social institutions and in other people. In this perspective, trust is based on norms and values shared by broad populations, usually at the national level (e.g. Putnam 1993; Fukuyama 1995). Russia is a low-trust country both generally and institutionally. The socialist experience had a profound impact in creating a low level of trust and shaping the present understanding of the concept of trust (Oleinik 2006; Rose 2000; Fukuyama 1995). Rose calls Russia an *antimodern* society in which official institutions have failed, leaving citizens no alternatives to influence their own affairs other than various unofficial institutions.

Many researchers who have emphasized the relationships between the economy and social networks have highlighted the importance of mutual trust between actors. Economic sociologist Mark Granovetter and the numerous researchers who have since done work in a similar vein have seen trust as a key factor in enabling economic networks to operate and thrive, and concrete personal relationships as essential in forming mutual trust (e.g. Granovetter 1985; Lorenz 1992; Maskell and Malmberg 1999). Several institutional economists, including Russian researchers, have emphasized the significant role that unofficial institutions, networks and personal relations play in the Russian economy (North 1997; Oleinik 2006; Radaev 1998; 2003).

The concept of trust has been applied in research on single-industry towns, where the social nucleus consists of relational networks between the production unit, owners and local workers. Both the success of the company and the livelihood of the local residents depend on the stability and undisturbed functioning of these local relations. The durability of these relations is far from self-evident: they are constantly contested and imbued with potential conflict, and thus face a recurrent threat of disintegration. Because the operation of a mill must be a routine activity in order to be economically and otherwise feasible, it does not allow daily negotiation of or struggles over the nature of local relations. The continuity of such relations, as well as industrial peace, is guaranteed by trust in actors and their willingness to follow mutually agreed or otherwise approved norms and conventions (see Phillimore and Bell 2005).

Tarmo Koskinen (1994) has studied Finnish mill communities and describes this relationship using the concept of social exchange. Such a system, and the relationships between actors in it, are characterized by deep trust, although Koskinen does not use this particular term. In his view, mills and the surrounding towns lived in a symbiotic relationship before the rise of the welfare state. For workers, the relationship meant that they had the security of a permanent job – usually extending over more than one generation – and that companies took care of the other needs of the workers by providing them with basic services and comfortable residential areas. For the companies, it meant confidence that they could obtain and keep the necessary skilled labour committed to working and living in the community.

In the Soviet Union, this symbiotic relationship between workers and managers was even stronger and supported by the communist ideology. In spite of an ideology emphasizing the power of the working collectives (workers and managers), the managers were able to make all the decisions, although these had to be based on state plans. The local authorities and the company managers were all state employees and ultimately governed by the Communist Party. Neither workers nor managers could choose where they worked. Mutual trust was well institutionalized, although it was based on personal relations. However, excessively close personal relationships constituted a threat to official institutions and led to corrupt relationships and the placing of personal gain before the fulfilment of official plans and aims (Nove 1977). The further the centre and its goals were, the more important were local relations and the local community. The spirit of belonging to the mill community was strong.

Trust does not emerge by itself; it is an outcome of mutual efforts to construct it. On the one hand, a company strengthens the relation of trust by investing in a community and making it a more attractive place to live for the employees than alternative locations; on the other, employees attain mill-specific skills, which cannot be replaced by an 'imported' labour force. Conversely, inept and imprudent operations may result in a lack of trust or even distrust between an enterprise and a community. The relation is by no means free of conflicts but very strained and highly susceptible to various kinds of disruptions.

In Russia and the Soviet Union pulp and paper mills have had the status of 'town-constituting enterprises,' because in many cases they were built in locations where large human settlements did not previously exist. In Soviet times, almost all the households in these mill communities were tied to the enterprises. The working population mostly laboured at the mills, and no strict dividing lines were drawn between the town administration and the mill management. The enterprise provided the entire community with health centres, kindergartens, housing, cultural centres, sports clubs, and other social and cultural services. Infrastructure such as district heating, a water supply and wastewater treatment was extended from the mills to cover the towns as well. There was neither municipal self-governance nor independent trade unions. (Holm-Hansen 2002, pp. 72–74; Yudakhin et al. 2002, pp. 12–14.)

Mill communities remained relatively unchanged in this respect, until the early 1990s. The transition to a market economy gave rise to new pressures on companies to restructure and modernize. The networks of the Soviet system started to loosen on all spatial scales and in all locations. Instead of fulfilling five-year plans and social obligations, the enterprises became economic actors striving to make a profit for their owners. In Russian mill towns this meant a new division between the towns and the enterprises, as many of the social tasks were transferred from the enterprise to the local administration. Generally speaking, the most rapid restructuring took place in the period 1993–1995 during the mass privatization of enterprises. The presidential decree of 10 January 1993 prohibited privatization of social and cultural capital together with the enterprises. According to the Privatization Programme, all the social and cultural 'objects' and responsibilities had to be transferred to municipal authorities within six months of the confirmation of the privatization plan of the company. Most municipalities, however, were not able to maintain municipal services. They later tried to solve the problem by privatizing municipal services through new companies.[3] These were not fully prepared for these tasks, however, and could not get payments from the citizens, who lived in an unmonetarized society – or *virtual economy*, as Gaddy and Ickes have called it (Gaddy and Ickes 1998).

After the first leap into privatization, the municipalization of social assets slowed, and President Putin and his administration started to emphasize the social responsibility of enterprises for the surrounding communities. 'Social responsibility' as used by the Putin administration refers to a better-managed transition to a market economy with a willingness on the part of companies to participate in developing the Russian economy and social system in cooperation with the government. Enterprises are now required to support the surrounding communities, regions and even the Russian Federation in their financial problems (Interview of G. Gref by Vera Sitnina 2003).

Poorly developed municipal self-governance is a burden on company-community relations. The federal legislature has tried to develop municipal self-governance through new laws giving more responsibility and power to the municipal level.[4] Local self-governance is a new institution, but difficulties in effecting reforms do not stem solely from the institution being new and difficult for people to understand. The most significant hindrance is the lack of revenues, which for the most part is due to tax and budgetary legislation. Most social expenses, such as schools and hospitals, fall to the local level, while tax revenues flow mainly to the federal government. The Law on the Financial Foundations of Local Self-Governance (27.12.2000) regulates how taxes are collected and what

---

3   The federal law on privatizing state and municipal property came into force on 30 November 2001.

4   A reform of municipal self-governance was to be started at the beginning of 2006. However, the new federal law on local self-governance was postponed, because municipalities were largely unprepared for the change.

the minimum level of revenues is that should remain on the local level.[5] However, there is no law regulating how the federation should redistribute revenues to the regions and municipalities. Redistribution of revenues is an effective financial means in the hands of the federal centre in levelling out economic differences between regions. Yet, the budgetary system puts local authorities at the mercy of the federal and regional authorities and the tax-paying companies in their area.

Business management and business cultures, which are important factors in the formation of relations between companies and workers, vary substantially between companies and between countries. Obviously, any transformation of a business culture will depend strongly on the character of the ownership of the local mill. Officially, Russian business culture is gradually transforming itself into one that is very close to the Western style of management and organization. Company legislation, which follows the models of the EU countries, suggests that Western-style corporate governance is the ultimate aim. However, observers who have studied the practices making up Russian business culture have maintained that Russian corporate governance and business culture will evolve into a unique model that reflects the country's traditions, values, and other cultural characteristics (McCarthy and Puffer 2002; Hendley, Murrell, Ryterman 2001; Oleinik 2005; Radaev 2003; Dolgopyatova (ed.) 2003).

McCarthy and Puffer see three cultural aspects as being the most important in this respect. Firstly, they emphasize the tradition of circumventing laws and other official rules, which is deeply rooted in the minds of the people. From the tsarist regimes through the Soviet era up to present times, it has been widely acceptable to ignore official rules that seem to make little sense to the people.[6] Secondly, mostly due to a deep feeling of mistrust engendered by the Soviet system, Russians tend to suspect individuals, groups, and organizations that fall outside their sphere of personal relationships. Thirdly, corporate governance is influenced by the Russian tradition of reliance on personal networks of trusted friends and colleagues to get things done (McCarthy and Puffer 2002). These results indicate a low level of general trust – especially where normative rules are concerned – in Russian society and a low level of trust in 'outsiders' or foreign actors; the findings also emphasize the importance of personal and family relations in the formation of trustworthy links.

In a mill community, there have to be trustful relations between the key actors and actor groups. In particular, sound relations between the owners, managers

---

5   According to the law on the financial foundation of municipal self-governance, a municipality should get at least 50 per cent of the income tax of the enterprises, 50 per cent of the property tax of the enterprises and likewise 50 per cent of the income tax paid by natural persons. The percentage of revenues received has to be negotiated with the region, which almost never agrees to give the municipality more than the minimum, which the law requires. This law will be repealed when the new Law on Municipal Self-Governance enters into force.

6   For more about the role of law and legislation in Russia, see, e.g., Berman 1996 and Nystén-Haarala 2001.

and workers are essential for the 'functioning' and success of the mills and communities. If there exists a deep and extensive feeling of distrust towards outsiders in Russian culture, as McCarthy and Puffer (2002) credibly argue, it may cause problems particularly for foreign investors. Distrust may cause and, as will be shown below, has caused difficulties even for Russian investors, because local residents may regard them as outsiders and intruders. Trust must be gained and constructed separately in each locality and situation in the contemporary Russian mill communities, and our case studies illustrate how difficult this can be and how differently it can be done.

The present owners of the Russian forest industry can be divided into four groups: the employees and managers of the mills; foreign investors; the Russian forest corporations; and the so-called oligarchs[7] (Kortelainen and Kotilainen 2003; 2006). These different owner groups represent rather distinct corporate governance and business cultures, which in turn strongly affect how the previously close relationship and trust between the enterprise and the community will develop in the changed circumstances. The differences become apparent if we compare two extreme models of constructing company-community relations, which we will call here the paternalistic model and the transnational model. After presenting these two models, we use them as bases for examining actual cases.

Two Models

The Paternalistic Model

Kondopoga is a town of 35,000 inhabitants in the Republic of Karelia that has grown up around a huge pulp and paper mill specializing in newsprint production. Kondopoga is the second largest town in Karelia and is situated 40 kilometres from Petrozavodsk, the republic capital. The enterprise is predominantly owned by the management and employees, with the German pulp and paper supplier Conrad Jacobson being a minority owner. There have not been extensive lay-offs at the mill. The enterprise has also continued to serve the local community with social infrastructure by providing a wide range of services and building new facilities for public use. No clear division exists between the enterprise and the town administration. This is rather rare in present-day Russia, because municipal property has officially been separated from mill property and most companies are trying to get rid of 'municipal services'. The company takes part in almost all fields of community life, invests in the local public services, and the taxes it pays account for about 80 per cent of all local taxes and 20 per cent of the taxes of the Republic of Karelia (interviews in 2004).

Although all services were transferred from the mill to the municipality in keeping with the 1993 presidential decree, the municipality constantly borrows

---

7   For more about the concept 'oligarch', see Chapter 2.

money from the enterprise, which seems to have no other alternative than to lend it in order to save the municipality from total bankruptcy. The salaries of municipal workers, such as teachers, doctors and nurses, represent 70 per cent of the municipal budget, and only 10 per cent of the budget is freely disposable (in 2004). According to the director of the budgetary committee of the municipality, who is also the director of social affairs of the mill, the law on the financial foundations of local self-governance makes calculating a real budget for the municipality impossible and forces the mill to support the municipality with continual loans. In this social and economic situation, the mill started to build schools and kindergartens again. The company has a construction department with 700 workers that takes care of both the production buildings and social facilities. It also provides heat and food to schools and kindergartens for free and has built a health care centre with a first-rate cardiology ward. Furthermore, the company provides financial assistance for local sports, cultural life and other activities for the residents and employees. According to the mill management, sports activities – especially fitness – are supported because they promote a healthy way of life for the youth of the town in particular and thus secure the availability of labour locally in the future (Itkonen and Stranius 2002, pp. 72–76; Melin and Blom 2002, 596–597; interviews in 2004).

It seems that Kondopoga is continuing the tradition of the 'town-constituting enterprise' that prevailed during the Soviet era. Even though the budgets of the municipality and the company are officially separate, the former is in practice totally dependent on the decisions of the latter in developing public services, which are officially company services but remain accessible to the local public. This symbiotic relationship is favourable for the municipality, because if the company did not invest directly in the community and lend money to it instead of putting the money into wages and dividends, most of its tax revenues would go to the regional or federal budget. If salaries were raised, only 35 per cent of the resulting rise in tax revenues would end up in the municipal budget.[8]

Much of this development owes to the ownership structure of the mill, in which the employees have an important role. At the same time, however, this situation is a consequence of the very strong influence of the general director of the mill, Vitaly Federmesser. He has been able to keep his position since the late stages of the Soviet era, and was the key person when the decisions were made about the mill's ownership and investments in social infrastructure. Kondopoga is a good example of how unofficial corporate governance is stronger and overrides the official rules. Shareholders' meetings are always arranged as the law prescribes, but no one would imagine criticizing the suggestions that the general director makes through the board of directors. Other shareholders (the employees and pensioners of the

---

8   According to the Law on the Financial Foundations of Local Self-Governance, municipalities should get at least 50 per cent of the local income taxes paid by individuals and enterprises. The parliament of Karelia, however, made a decision reducing the share of the municipalities to 35 per cent and increasing the share of the region correspondingly (interviews in Kondopoga and Petrozavodsk 2004). Cf. footnote 5.

company) sometimes reluctantly accept the reasoning of the general director, but cannot imagine there being any other choice.

Finnish researcher Harri Melin, who examined social change in Kondopoga in a 2005 study, calls the local system 'new paternalism'. Despite its benefits, this 'paternalistic model' is not without problems. The salaries of the employees are reported to be relatively high but still about one-third lower than in other large paper mills in Russia. Dependency on the enterprise, its employment, housing and services restricts peoples' lives, opportunities and even freedom of expression. According to Melin and Blom (2002), local residents are not eager to discuss problematic topics, because they are afraid of trouble at work. Criticism seems to be growing gradually and becoming more open, because in our interviews in the autumn of 2004 representatives of both the trade union and management were ready to analyze the pros and cons of the situation without any fear for their jobs. Employees were reluctantly ready to accept the explanation that higher salaries would cause inflation and make the standard of living of other inhabitants too low compared to that of the mill workers. However, everyone was ready to both admit and accept that decisions are made in accordance with the ideas of Vitaly Federmesser.

The huge investments in the built environment in Kondopoga have also prompted some criticism. The new high-quality cardiology centre has not been criticized, but the luxurious indoor ice stadium and culture palace, both featuring marble surfaces, have prompted some discussion, as people are suffering from a housing shortage and complaining of expensive housing costs. On his visit to Kondopoga, President Putin wondered why part of the money spent on these facilities had not been spent on housing production (Itkonen 2003). The decisions were Federmesser's and it seems that when he retires in the near future, now hidden tensions will surface, with the company either ending up in a long struggle to establish its new direction or gradually becoming more market economy oriented. Federmesser's announcements in the press, just before our interviews in 2004, about considering retiring and moving to Germany seemed to confuse the whole town.

Vitaly Federmesser has built an efficient network of good relations with the regional and federal officials. President (now Prime Minister) Putin's family and Federmesser's family visit one another. When the parliament of the Karelian Republic made a decision to sell its shares in several Karelian enterprises, including its 20 per cent share in the Kondopoga paper and pulp mill, Federmesser agreed to buy the cardboard mill in Suojärvi and invested 20 million roubles in modernizing the mill in order to make the parliament rescind its decision to sell its share of the Kondopoga mill. However, Federmesser was able to refuse to save another small mill in Läskelä, appealing to the excessively high cost of doing so. The mill management in Kondopoga proudly explained that Federmesser can have his way in negotiations with regional authorities. The leadership of the Republic of Karelia seems to admire Federmesser and his policies. Valery A. Shlyamin, former foreign relations minister of the republic, presents the Kondopoga mill as an ideal model of doing business in Karelia and mentions Federmesser as the person responsible for the success of the mill (Shlyamin 2002, 56).

The Transnational Model

The largest of the completely foreign-owned mills in Russia are the Svetogorsk pulp and paper mill in Leningrad Region and the Syktyvkar pulp and paper mill in the Republic of Komi. Svetogorsk is at present owned by an American company, International Paper,[9] whereas the Syktyvkar mill belongs to the Austrian company Neusiedler, which is part of Mondi Europe, owned, in turn, by a multinational corporation, Anglo-American. Both mills have been said to be among the most successful and advanced pulp and paper producers in Russia (Butrin 2002).

Svetogorsk is a single-industry town of 15,000 inhabitants located in Leningrad Region near the border between Finland and Russia. The local pulp and paper mill has been owned by foreign companies since the mid-1990s, having had a Swedish owner prior to its present American one. Svetogorsk is in many ways an opposite example to Kondopoga. The mill there has been characterized by a strong decrease in the number of employees as well as a tendency to transfer as many social assets from the mill to the municipality as possible (see Bolotova and Vorobiov 2002). The town is still very much dependent on the mill, the site of almost all its industrial jobs. From 1989 to 1999, the drop in the number of workers at the mill was more than 50 per cent. The dramatic change can be explained by not only more efficiently organized production but also the municipalization of social assets.[10]

In Svetogorsk the loosening of the ties between the mill and the locality and foreign direct investment in the mills have been simultaneous processes. Foreign investment has increased the pressures for municipalization. The mill and the town have been separated since 1997, and each is supposed to take care of its own particular responsibilities. Nevertheless, the enterprise still provides, among other things, the heating of the flats in the locality, a fire brigade, and water and sewage systems.

In spite of lay-offs, the enterprise seems to have gained the trust of the employees and local community. The reasons for this are partly economic, as the mill is able to provide its workers with regular and relatively high salaries. Average monthly wages in Svetogorsk are about 50 per cent higher than elsewhere in the surrounding Vyborg district. Only one district in Leningrad Region, Kirishskiy, which has heavy industrial production, has higher average salary levels (Kortelainen and Kotilainen 2003, 397). International Paper has built a high-level health centre for its workers, but the steadfast aim of the company management has been to get rid of the rest of the community responsibilities and reduce the number of employees further (e.g. Brown-Humes 2001). At the same time, there have been efforts to

---

9   In March 2008 it was announced that International Paper and Russian Ilim Pulp Enterprise have started to form a Russian holding company.

10   For more about Svetogorsk, see Kotilainen 2004.

increase local self-government in Svetogorsk. Previously, the town was part of the Vyborg district, but in 1995 it became a municipal entity in its own right, with the boundaries of the town established in 1998. One motive for the change was to keep taxes on the local level (Kortelainen and Kotilainen 2003).

Svetogorsk can be seen as a case of the transnational model as regards the relationship between the enterprise and the local community. The locality has become included in the transnational networks of International Paper, a company that aims to implement similar concepts wherever it has production facilities. Thus, it has sought to transfer its own work and business culture to Svetogorsk as well. Obviously, the results of this transfer process differ from place to place due to the varying national and local contexts.

Alternative Company-Community Relations

Oligarchs and Distrust

Ownership arrangements have affected communities and their development in various ways. Changes of ownership are always extremely sensitive situations, ones that easily create conflicts between owners and employees. Some of the new owners have managed to cause confusion and distrust in communities, and especially outside business people and oligarchs have prompted suspicion. In recent years, a highly concentrated corporate structure has emerged in Russia. A small number of cash-rich exporters have been purchasing the assets of various industries across the country. With the most lucrative sectors – the oil, metal and automobile industries – already divided, the forest industry became a prime object for carving up (Boone 2001; Latynina 2002; Anna-Maija Matilainen describes the methods of forest industry takeovers in chapter 6 of this volume). This has often turned out to be the most conflict-ridden type of ownership.

Most companies were privatized and sold to their managers and workers in the first wave of privatization (1992–1994), and many have later been sold on to new owners. There were also cases in which either the management or the new owners only aimed at squeezing immediate profits out of the company before running it down. Sovietskiy mill (Leningrad Region) was a typical example of a carving-up policy of the early and mid-1990s. After privatization, the new shareholders and management bought the workers out, established almost full control of ownership and then sold the factory. The company that formally obtained ownership rights was a typical 'dummy' company which provided false information. The workers saw that the company's purpose was not to develop the mill but to play the market and provide immediate profits for the business people who owned it. There was a very real threat that the owners would close the mill, and sell the machinery, with the workers losing their jobs. The situation became extremely critical when the workers seized the mill and prevented the new owners from entering it. This situation caused a major conflict where even firearms were involved (Kolganov

1999; Davydova 2002). Sovietskiy is an example of total unwillingness to even try to create trust, an attitude quite common in the early 1990s. In friendly takeovers, when a successful company is willing to invest in and run the mill, the prerequisites for building trust are better than in hostile takeovers or carving-up attempts.

Nowadays hostile takeovers are a nuisance in the Russian economy. The most famous of the oligarchs operating in the forest industry is Oleg Deripaska and his holding company Basic Element, which has ownership disputes with various enterprises. Basic Element's main adversary has been Ilim Pulp Enterprise, a domestic forest corporation which has its headquarters in St. Petersburg and controls three major pulp and paper mills (Kobyzev 2001; Butrin 2002). These companies have struggled over the ownership of Ilim Pulp's mills. Despite several takeover attempts and lawsuits, Ilim Pulp has managed to keep control over the mills (see chapter 6).

During the ownership struggle for the Kotlas mill in summer 2002, the media reported that both of the competing companies had armed troops standing by, Ilim Pulp's people being in the mill and the opponents in the vicinity. Local workers and residents strongly supported the present owner, Ilim Pulp, and openly resisted the attempt to take over the mill, which was considered a hostile act carried out by outsiders. The workers held a major demonstration where speeches were given and the will to defend the mill was emphasized. They described the new owners as 'bellicose groups from big business' that would start 'seizing, remaking, and plundering' the mill and stated emphatically that they were the only ones who could stop the owners (Bratkov 2002). No armed conflict ever erupted, and after several lawsuits Ilim Pulp made a deal with Basic Element (in spring 2004) on putting an end to takeover attempts. The situation is interesting from the point of view of the construction of trust, because only a couple of years earlier Ilim Pulp was a new owner in the community. It had to put a great deal of effort and money into local relations before it could earn the trust of workers and other residents in the community.

Problems Encountered by Foreign Investors

Foreign investors, in particular large forest corporations, have shown increasing interest in Russian pulp and paper mills, and some mills have already been bought by foreign companies. The failure of the Swedish company AssiDomän with the Segezha mill serves as an example of how a lack of trust between a foreign company and Russian stakeholders as well as inept measures can bring down a large-scale investment project. AssiDomän acquired the majority of shares in Segezhabumprom in 1996, a company running a large pulp and paper mill in the Republic of Karelia. The next year the company appointed its own representative as managing director but simultaneously found itself in a serious dispute with federal and Karelian government officials. There were disagreements over taxes, pension liabilities, cutting rights and, ultimately, the legitimacy of the company's ownership. The mill was closed down and the community left without heating for months, which gave rise to dissatisfaction among the residents. Finally, the Swedish company withdrew from Segezha and lost its entire investment, which totalled tens of millions of US dollars (Kauppalehti 1998; Brakonier 1997).

In this case, the problems stemmed not so much from a failure in building up a successful local enterprise-community relationship as from distrust between AssiDomän and regional officials. Some observers have criticized the Swedish company for going into Russia without knowing the real nature of the local culture and circumstances. The Swedish managing director did not speak Russian and he was unable to establish the necessary relations with powerful actors in the government. The Karelian Republic turned out to have a difficult economic atmosphere, where foreign investors were shunned.[11] After the failure in Segezha, AssiDomän turned its eyes to other areas and has since succeeded in projects in St. Petersburg, a development which has reportedly provided a very favourable economic environment for foreign direct investments (*Kauppalehti* 1998). The Segezha mill's current Russian owners, who come from Moscow-based banking circles, are now seriously investing in building trust with the local community and logging villages in the mill's logging areas. The owners seem to be pursuing a policy that is quite similar to the trust building of Ilim Pulp (interviews in Segezha and Padani in 2006).

A different example of the problems foreign investors face is what happened at the Arkhangel'sk pulp and paper mill in Novodvinsk. The company previously belonged to the holding group Titan, owned by an Arkhangel'sk-based oligarch, Kurpchak. Austrian Pulp Mill Holding GmbH bought a majority of the shares in the mill in May 2003 in a 'friendly' takeover. Oleg Deripaska's Basic Element almost immediately sought control of the Arkhangel'sk mill. In summer 2005, Basic Element owned 34 per cent and Austrian Pulp Mill Holding GmbH 65 per cent after Wilfried Heinzel AG sold its shares to Deripaska. Deripaska has since tried to win over the workers and the community, for example, by claiming that he has better social and ecological plans – including higher salaries for the mill than the present owners pay. Basic Element has apparently failed in its attempts to create and utilize the presumed distrust of the foreign owners (cf. McCarthy and Puffer 2002). In our interviews, the representatives of the trade union did not seem to have a negative attitude towards foreign owners. On the contrary, foreign investments seemed to represent hope for a better future. It may well be that Russian employees tend to be loyal to their employer, probably because of the 'we and our enterprise' spirit inherited from the Soviet times. Mistrust towards the present owners is therefore usually more a result of unsuccessful management than suspicion towards foreigners.

Due to the lengthy ownership conflict, the foreign owners have been unable to restructure, let alone lay off labour at their mills. They have ended up in a situation

---

11   Former Foreign Minister Shlyamin, who admires the Kondopoga model in his book, criticizes the Swedish owners of the Segezha mill exceptionally heavily for pursuing a policy that was not suitable for Russian business (Shlyamin 2002, 61).

which any foreign investor would like to avoid in Russia. Hostile takeover attempts keep them hostage in Novodvinsk without being able to concentrate on developing business. Employees, however, did not seem to be afraid that foreign ownership would lead to a loss of jobs. In fact, they seemed to have overly high hopes for international support from their co-workers in the company abroad. One reason for their acceptance of foreign ownership may also be that the owners have not been able to make a lot of changes and have not laid off workers. The town of Novodvinsk is dependent on the huge paper and pulp mill. Even though Novodvinsk is quite near the city of Arkhangel'sk, the city would likely have difficulties in absorbing a large number of unemployed pulp and paper mill workers from a nearby town. (For more on the ownership conflict, see chapter 6.)

Russian Market Economy Models

The Solombala pulp and paper mill, like Kondopoga, is owned by employees and managers but is smaller and not situated in a single-industry town; it lies on the outskirts of the city of Arkhangel'sk, which has 358,500 inhabitants. The Solombala pulp and paper mill has strived to concentrate on business only. Such a policy has been possible, because the company pays taxes to the city of Arkhangel'sk along with many other tax-paying enterprises. The company sponsors the local bandy team Vodnik and other sports and cultural activities, acting quite similarly to any Western sponsor. However, the mill also participates in the city's and the region's social programmes just like any other successful enterprise in the city is expected to do. Sponsoring municipal programmes reflects attention to public relations as well as social responsibility. In Russia enterprises are not as free to choose how they show social responsibility as companies in developed Western market economies are. Staying out of local or national programmes would harm the reputation of the company, because companies are the only units that are assumed to have money and to be able to pay the costs of social programmes. Companies also have to support schools and invalids, because the tax revenues of municipalities are completely inadequate for maintaining ordinary municipal services.

Even though the managers of Solombala firmly adhere to their market economy policy, restructuring has been quite slow. Just like the Kondopoga mill, the Solombala company has to collect the money for new machinery before it can buy it, the reason being the high interest rates on bank loans.[12] Restructuring has also been hindered by claims of malfeasance during the first wave of privatization (1992–1994), when the company was privatized and sold to the managers and workers. The cases to be prosecuted were dropped in court (in 2004) because they were barred by the statute of limitations. According to the company management,

---

12   There is a difference between the Kondopoga and Solombala interviews in 2004 and the Segezha interviews in 2006. The Segezha management said that bank loans were already a reasonable option for restructuring purposes. Segezha, however, is owned by the Bank of Moscow.

these court decisions opened the way for investments and gradual restructuring. The market-based policy seems to have the support of the shareholders and managers alike, many of whom are major shareholders. Trust, however, has to be maintained, especially in the city, which has a constant need for support in its social tasks.

The most acute threat at present, however, is takeover attempts, which may be directed at any successful company on the market. The director of legal affairs of the company is a member of the regional parliament and several managers are members of the municipal council. The interests of the company and business generally are well represented at the regional level, with many local entrepreneurs and business managers serving as members in the regional parliament. However, it seems that any company should have good relations with the federal centre, too, because there is currently a strong trend at the federal level to 'uncover' corrupt regional power elites. Satisfied shareholders, as well as successful lobbying and the good relationships of the top management, also seems to have caused envious attitudes among the inhabitants of Arkhangel'sk. Combined with the federal battle against regional cliques, this situation causes a potential danger for the Solombala mill, which has to strike a balance that maintains trust with various stakeholder groups.[13]

The Russian corporation Ilim Pulp[14] has tried to imitate Western companies in its corporate governance by putting an emphasis on social and environmental responsibility in the same manner as its Western 'role models' (see Ilim Pulp 2004). However, at the same time it continues some national traditions. When the company bought its mills in the 1990s, the workers in each community were highly suspicious of the new owner. The company tried to improve the situation by paying the wages that were in arrears, raising salaries, implementing new staff policies and carrying out media campaigns. Following the example of Western companies, it established a human resources and public relations department. The company also invested large sums of money in social services in the mill communities, a more traditional measure (Kobyzev 2001, 170). With these measures it gradually earned the trust and support of the local people, which was visible in the above-mentioned ownership conflict in Kotlas.

The new owners of Segezha pulp and paper mill have adopted a strategy similar to that of the Ilim Pulp company. Compared to the established owners of Solombala, who only have to maintain the trust of the employees, the new owners have to prove that they are capable of running the company. Solombala has been quite cautious in its actions and tried to keep a low profile. Segezha, on the other hand, has accepted that it has to support the local community, but its ideology

---

resembles more the Solombala than the Kondopoga model. Segezha is trying to become a Western-style market-oriented company with high social responsibility. It is striving to obtain FSC forest certification with more responsible forestry and more attention to the living conditions in the logging communities and old villages inhabited by ethnic Karelians in their logging areas. With these efforts, Segezha is trying to combine market success and social responsibility (interviews at Segezha mill and Padani village in 2006).

Conclusions

The above analysis shows that there are large differences in the construction of trust between different mill communities. Most of the successful pulp and paper mills have received new owners since the early 1990s. The emergence of new owners first raised suspicions among the local public. Oligarchs prompted especially strong negative reactions, because people were afraid that they would only rob the property and then leave the community on its own. Even comparatively reliable companies have had to build up and invest in trustful relations in mill towns. The above examples illustrate two extreme models that are in use in Russia and their mixed variations. Kondopoga represents the paternalistic model, while Svetogorsk embodies the transnational model. Most of the enterprises fall in between the two. Solombala is trying to implement a market economy model, but cannot leave behind the burden of its social responsibilities, which provide services that the municipality needs from its enterprises. The Solombala mill does not actively have to build trust with its employees, but it has to strike a balance between its own needs and those of the City of Arkhangel'sk. The partly foreign-owned Arkhangel'sk paper and pulp mill would probably strive to adhere to the transnational model but has been hindered by takeover attempts and the strong dependency of the inhabitants of the city of Novodvinsk on the company. The new Russian-owned Ilim Pulp Enterprise and the Segezha Company are putting a great deal of effort into building trust with the local communities.

Locally specific models are part of the broader development issues of transition economies. Economist Clem Tisdell (2001) has pointed out that there exist two kinds of ideal models of capitalism. The first is *laissez-faire*, which puts an emphasis on complete freedom of the markets. The second is a social market economy, which emphasizes social responsibility and justice. The rapid transition to a market economy and the declining importance of nation-states associated with economic globalization have strengthened the position of the *laissez-faire* ideal in Russia as well as other transition economies. Putin's regime has, however, encouraged taking a few steps back with its demands for social responsibility.

The above studies of mill communities show, however, that there have existed different variants of a local social market economy in the communities around successful mills even before and independent of Putin's emphasis on social responsibility. The *laissez-faire* market economy and weak central government are not able to produce general trust which would guarantee stable economic development. In many of the Western countries, enterprises have not had to worry about local trust, because the welfare state has taken care of trust building on the national level. In contemporary Russia, each enterprise has to create trustworthy relationships locally by itself and in its own way. Economic sectors such as pulp and paper require large numbers of skilled workers and production is not possible without a permanent workforce. Enterprises have themselves had to create sustainable social and economic circumstances in mill communities. In some communities, such as Kondopoga, the enterprise has kept most of the tasks in its own hands, partly reflecting an ideology of social responsibility and justice inherited from the Soviet times. In some other cases (e.g., Svetogorsk), decisions are based on economic considerations. Enterprises aim at acting as indirect sponsors and transferring responsibilities to the local government to the extent possible.

The role of foreign capital is increasing and many Russian producers tend to adopt Western ideas. It is worth remembering, however, that internationalization does not equate with homogenization. There still exist considerable differences in how Russian and Western firms operate. Furthermore, there are immense differences between management and business cultures in different Western countries and these affect how they work in Russia. While trying to enforce their own transnational management and organization systems in all the countries and localities in which they operate, they have to adapt themselves to national and local circumstances. It is also evident that Russian companies applying the paternalistic model have to adjust themselves to increasing competition. It is most probable that companies like Kondopoga have to cut their expenses, which would render the bond between the enterprise and community thinner. There has already been discussion in Kondopoga whether the focus should be redirected from services to better salaries. This may well happen when the creator of the Kondopoga model, general director Vitaly Federmesser, retires and loosens his hold on the company.

Most probably the new model of the Russian mill community will be a hybrid of the paternalistic and transnational models – a more or less socially responsible model of the Russian type depending on local circumstances. Strong cultural and social traditions force foreign companies to adjust their activities accordingly. The laws of the market will force the Russian companies to cut costs in the long run, which probably will result in declining jobs and diminishing service provision.

Russian cultural and social traditions may also hinder foreign investments. Since public relations are a demanding task for Russians, who know their culture, foreigners find it even more difficult to balance between the requirements of regional and federal authorities and keep good but not too close (and corrupt) personal relations with them. Creating trust in the local community and with the employees may even be easier than maintaining good relations with authorities. Employees are ready to welcome foreign owners who can show that they are seriously planning to invest in the company. Russians' suspicious attitude towards foreign owners has started to fade into history. However, distrust and a Soviet-style dictating mentality may still be found among Russian authorities.

References

Berman, H.J. (1996), 'The Struggle for Law in Post-Soviet Russia', pp. 41–55 in Sajo, A. (ed.) *Western Rights? Post-Communist Application* (The Netherlands: Kluwer).

Bolotova, A. and Vorobiov, D. (2002), 'Local case study II: Svetogorsk Pulp and Paper Mill', pp. 97–108 in Kortelainen and Kotilainen (eds).

Boone, P. (2001), *Turn Around in Russia's Corporate Sector*. Presentation held at the Russia Ten Years After conference. Moscow, 8–9 June 2001. http://www. ceip.org/files/programs/russia/tenyears/panel2.htm.

Brakonier, P. (1997), '5500 tvingas bort från AssiDomäns ryska fabrik', (5 500 are Forced Away from AssiDomän's Russian Mill) *Dagens industri*, 8 mars 1997. (Swedish)

Bratkov, V. (2002), 'Epidemic of Proletarian Mutinies in the Arkhangel'sk Region', *Pravda.ru* 10.7.2002.

Brown-Humes, C. (2001), 'Western Capitalism and Small Russian Town', *Financial Times* (UK) 11.6.2001.

Butrin, D. (2002). Who Owns Russia: 'The Timber Industry. The Sector Today', *The Russia Journal*, July 19. – 25. 2002.

Dasgupta, P. and Serageldin, I. (eds) (2000), *Social Capital. A Multifaceted Perspective* (Washington: The World Bank).

Dolgopyatova, T.G. (ed.) (2003), *Russian Industry: Institutional Development* (Moscow: State University Higher School of Economics).

Fukuyama, F. (1995), *Trust: The Social Virtues and the Creation of Prosperity* (Hammondsworth: Penguin).

Gaddy, C.G. and Ickes, B.W. (1998), 'Beyond the Bailout: Time to Face Reality about Russia's Virtual Economy', *Foreign Affairs* 77: 53–67.

Granovetter, M. (1985), 'Economic Action and Social Structure: The Problem of Embeddedness', *American Journal of Sociology* 91:3, 481–510.

Hendley, K., Ickes, B.W., Murrell, P. and Ryterman, R. (1997), 'Observations of Land Use of Russian Enterprises', *Post-Soviet Affairs* 13, 19–41.

Holm-Hansen, J. (2002), 'Institutions for Environmental Protection in an Industrial Town of Northwest Russia', pp. 71–109 in Jørn Holm-Hansen (ed.).

Holm-Hansen, J. (ed.) (2002), *Environment as an Issue in a Russian Town*, NIBR-Report 2002:11. Oslo.

Ilim Pulp (2004), Social Report. (http://www.ilimpulp.ru/reports/social_report. pdf). Ilim Pulp, St.Petersburg.

Itkonen, A. (2003), 'Tehtaan valoisa varjo', (The Light Shadow of the Mill) *Karjalainen* 23.4.2003. (in Finnish)

Itkonen, H. and Stranius, P. (2002), '*Liikettä Venäjän Karjalassa*', (Movement in Russian Karelia) Publications of the Karelian Research Institute 137, University of Joensuu. (in Finnish)

Kauppalehti (1998), 'Assidomänin Segeza-seikkailu oli virheiden summa. (Assidomän's Segezha Adventure was a Sum of Mistakes) Optio Special'. *Kauppalehti* 6/1998: 58–62. (in Finnish)

Kobyzev, S.B. (2001), 'Internationalization of a Russian Pulp and Paper Company – JSC Ilim Pulp Enterprise', pp. 62–77 in Liuhto, K. (ed.).

Kolganov, A. (1999), 'The Labor Movement in Russia: Changes Ahead', PRISM 5:19 (November 19,1999)http://www.jamestown.org/publications_details.php?volume_id=6&ise_id=406&article_id=3702 (accessed 11.5.2006).

Kortelainen, J. and Kotilanen J. (eds) (2002), *Environmental Transformations in the Russian Forest Industry: Key Actors and Local Developments*. University of Joensuu, Publication of Karelian Institute No. 136.

Kortelainen, J. and Kotilainen, J. (2003), 'Ownership Changes and Local Development in the Russian Pulp and Paper Industry', *Eurasian Geography and Economics* 44:5, 384–402.

Kortelainen, J. and Kotilainen, J. (2006), *Contested Environments and Investments in Russian Woodland Communities* (Helsinki: Kikimora Publications).

Koskinen, T. (1994), '*Elämää yhteisössä*', (Life in a Community) Publications of the University of Vaasa. Research 176, Sociology 9. (in Finnish)

Kotilainen, J. (2004), 'Shifting between the East and the West, Switching between Scales', pp. 133–156 in Lehtinen, A., Sæther, B. and Donner-Amnell, J. (eds).

Latynina, Y. (2002), 'Forest Industry: Ripe for Carving Up', *Moscow Times*, July 17 2002.

Lehtinen, A., Sæther, B. and Donner-Amnell, J. (eds) (2004), *Politics of Forests: Northern Forest-industrial Regimes in the Age of Globalization* (Aldershot: Ashgate).

Liuhto, K. (ed.) (2001), *East Goes West – Internationalization of Eastern Enterprises*, Lappeenranta University of Technology, Studies in Industrial Engineering and Mangement No 14, (Lappeenranta: University of Lappeenranta).

Lorenz, E.H. (1992) 'Trust, Community and Cooperation: Towards a Theory of Industrial Districts', pp. 195–204 in M. Storper and A.J. Scott (eds).

McCarthy, D. and Puffer, S. (2002), 'Corporate Governance in Russia: Towards a European, US, or Russian Model?', *European Management Journal* 20:6, 630–640.

Maskell, P. and Malmberg, A. (1999), 'The Competitiveness of Firms and Regions. 'Ubiquitification' and the Importance of Localized Learning', *European Urban and Regional Studies* 6:1, 9–25.

Melin, H. (2005), 'Towards New Paternalism in Kondopoga', pp. 61–75 in Melin, H. (ed.).

Melin, H. (ed.) (2005), *Social Structure, Public Space and Civil Society in Karelia* (Helsinki: Kikimora Publications) B34.

Melin, H. and Blom, R. (2002), 'Toisenlainen Venäjä', (Another kind of Russia) *Yhteiskuntapolitiikka*, 67:6, 595–599. (in Finnish)

Melin, H. and Nikula, J. (2005), 'Social Structure of Karelia', pp. 145–151 in Melin, H. (ed.).

North, D.C. (1997), The Contribution of the New Institutional Economics in Understanding of the Transition Problem. The United Nations University, World Insitute for Development Economics Research (UNU/WIDER) Annual Lecture, 7 March, Washington University, St. Louis.

Nove, A. (1977), *The Soviet Economic System* (London: Allen and Unwin).

Nystén-Haarala, S. (2001), *Russian Law in Transition* (Helsinki: Kikimora Publications) B21.

Oleinik, A. (2005), *Konstitutsiya rossiiskogo rynka*, (The Constitution of the Russian Market), pp. 395–437 in Oleinik, A. (ed.). (in Russian)

Oleinik, A. (ed.) (2005), *Institutsional'naya ekonomika* (Institutional Economics) (Moskva: Infra). (in Russian)

Phillimore, P. and Bell, P. (2005), 'Trust and Risk in a German Chemical Town', *ETHNOS* 70:3, 311–334.

Putnam, R.D. (1993), 'The Prosperous Community: Social Capital and Public Life', *The American Prospect* 4:13, 35–42.

Radaev, V.V. (1998), 'Korruptsiya i formirovanie rossiiskih rynkov: otnosheniya chinovnikov i predprinimatelei', (Corruption and Structuring the Russian Market: Relations Between Officials and Entrepreneurs) *Mir Rossii*, No. 3, pp. 59–70. (in Russian)

Radaev V.V. (2003), *Sotsiologiya rynkov: k formirovaniyu novogo napravleniya.* (Sociology of Markets: Developing a New Direction) (Moscow: Moscow State University and Moscow School of Economics). (in Russian)

Rose, R. (2000), 'Getting Things Done in an Antimodern Society: Social Capital Networks in Russia', pp. 147–171 in Dasgupta, P. and Serageldin, I. (eds).

Shlyamin, V.A. (2002), *Rossiya v severnom izmerenii.* (Russia in the Northern Dimension). (Petrozavodsk: Petrozavodsk State University Press).

Sitnina, V. (2003), 'Interview of Economic Development and Trade Minister German Gref', *Vremya novostei*, 22 December 2003. (in Russian)

Storper, M. and Scott, A.J. (eds) (1992), *Pathways to Industrialisation and Regional Development* (London: Routledge).

Tisdell, C. (2001), 'Transitional Economies and Economic Globalisation. Social and Environmental Consequences', *International Journal of Social Economics* 28:5/6, 577–590.

Yudakhin, F., Davidov, A., Ivanov, A. and Holm-Hansen, J. (2002), 'A Russian-Type Single Enterprise Town: The Case of Koriashma', pp. 11–24 in Holm-Hansen, J. (ed.).

Chapter 9

Conflict as a Form of Governance: The Market Campaign to Save the Karelian Forests

Maria Tysiachniouk

Introduction

Globalization is changing the traditional ideas and conceptions of the world order and overall global governance through multiple stakeholder involvement in both public and private policy making. The compression of time and space, the increased density of global networks, as well as the increased speed of information dissemination, flows of capital, goods and technologies lead to transnationalization of the global economy and have dramatically changed global patterns of production and consumption. In these processes, NGOs are essential in setting agendas for innovative global policies and private regimes (Meidinger 2003).

Although the role of NGOs and environmental movements has been addressed by many globalization scholars, the literature contains very few examples of efforts to apply a contemporary sociological understanding of time-space compression and a network-and-flow perspective to the analysis of NGO cross-border activities. In this paper, I analyze a deliberately initiated transnational conflict, 'a market campaign', which NGO networks use all over the world in order to force business actors to change their practices. A market campaign is usually related to the behaviour of multinational companies in both producing and consuming countries and pertains to the process of production and management of natural resources. The campaigns are typically initiated against corporations with a highly visible brand in cases where their operations violate internationally accepted rights of workers or destroy the environment and natural resources. Such conflicts are commonly organized by civil society actors to change the behaviour of not only forest-producing companies, but also companies in industries such as oil, textiles, bananas, and coffee. In this type of conflict, market forces and consumer preferences are used by campaign organizers to influence companies' behaviour.

The impact of NGOs on the practices of commercial firms during the past two decades has been so significant that some writers (Bernstein 2001; Cashore et al. 2004) have identified them as a new social force, which they call 'market-driven governance systems'. I look at NGO-led market campaigns, despite their being conflicts, as another form of governance that makes business practices more environmentally friendly and helps promote and construct environmentally friendly niche markets. In the same vein, Meidinger (2003a; 2003b) uses the case of forest certification programmes to argue that global civil society organizations act parallel to government and market institutions that create transnational norms and global governance systems. The outcome of their market-constructing role is contingent on its timing, socio-political context and interaction with various stakeholders in a particular field.

I am using as an example the market campaign that was initiated by NGO networks to save old-growth forests in the Republic of Karelia in the Russian Federation. Russia is interesting in this respect, as it represents a rapid transition from an extremely static form of governance to multi-stakeholder governance of natural resources with significant involvement of transnational actors.

The chapter is based on field research conducted between 2001 and 2006, in the late stages of the conflict. Qualitative methods, such as semi-structural interviews (N=52) and participant observation were used. The interviewees comprised NGO representatives in both Moscow and Karelia, representatives of governments at different levels in Petrozavodsk, Kostomuksha (Kostamus) and Kalevala (Uhtua), and stakeholders of Kalevala National Park, including people living in the park (see the map at the beginning of the book).

Theoretical Approaches

I analyze the consumer boycott organized by NGOs using the contemporary sociological understanding of space and time. Communication technologies allow sociologists to make the distinction between the space of place and the space of flow. The Internet and other technologies create special spaces of flow which allow actors to ignore geographical distance and to interact simultaneously (Castells 1996, 412). Unlike Castells, I do not draw a strict distinction between the space of place and the space of flow. I understand the transnational space as being more extensive then the Internet or the spaces provided by other communication technologies. I see it as a space of decision making that is not permanently localized in a concrete geographical setting. International conferences, UN or European Union bodies, NGO networks, business networks and corporate headquarters all belong to transnational spaces. By the space of place I understand a concrete geographical setting in which actions are taken by concrete actors. The actors involved are usually not single organizations, but entire networks that extend both throughout the country and across borders; thus, there is always a span from concrete localities to transnational spaces.

The new understanding of space triggers a new understanding of time (Urry 2000, 107). Communication technologies make information available both instantaneously and simultaneously across the globe. However, the day-to-day practices of people continue to occur in clock time. Both times co-exist and are relevant for contemporary sociological analysis. Transnational networks operate in both clock time and timeless time.

In the course of a market campaign, NGOs operate both in spaces of place and in transnational spaces, although their role is different in each of these spaces. One of the focuses of this paper will be the tensions and interactions between the space of place and the transnational spaces that appeared during the market campaign analyzed.

In conceptualizing the role of NGO networks in the space of place and in transnational spaces, I draw on the theory of organizational isomorphism (DiMaggio and Powell 1983) and path dependence (Arthur 1994). The theory of organizational isomorphism is based on three types of institutional influences to adopt an organizational innovation: coercive, mimetic, and normative (DiMaggio and Powell 1983). According to the literature, coercive pressures are created by actors (usually government agencies) who have the legal authority to oversee and regulate important aspects of human or organizational behaviour. In my work, it is not governments, but non-state actors and especially NGOs that played such a coercive role in discouraging the logging of old-growth forests through a consumer boycott.

Normative influences are created by actors whose social status puts them in a position to endorse and bestow legitimacy on a particular standard or practice, and thus sway other actors to adopt it. The characteristic feature of such systems is their reliance on normative appeals to consumer preferences and their use of market mechanisms to bring about the adoption of environment-friendly standards and technologies by commercial firms. Therefore, for the purpose of the present analysis, I treat these conceptualizations of the role of NGOs as a special case of normative isomorphic pressures exerted by influential, organized global NGO networks. NGOs use market campaign to build new values and, on that basis, consumer preferences. Simultaneously, other NGOs create alternative supra-state regulatory systems, such as FSC certification, which are also based on new norms. To force multinational corporations to take paths to certification, more radical NGOs organize consumer boycotts and less radical ones promote certification systems using both coercive and normative adaptation systems.

Mimetic tendencies result from a desire to succeed in a competitive environment, which prompts actors to adopt solutions that have given their competitors an advantage. Decision makers find it beneficial to adopt solutions developed and tried by others. Due to this mimetic form of adaptation, more and more companies have taken the FSC certification path.

Another incentive to adopt a new practice results from changes in the structure of transaction costs. Since such costs usually constitute a significant share of the overall costs incurred by organizations in attaining their goals, eliminating or reducing the costs is a powerful economic incentive for rational actors. Transaction costs are often reduced by adopting standards and solutions developed by other organizations in a particular field or geographic area. This increases the likelihood that a newcomer to that field or area will adopt the standards and practices developed by organizations already operating there, a process known as *path dependence* (Arthur 1994).

Transaction cost considerations are also the key factor affecting the supply side of market-driven governance systems (normative influences affect the demand side). Consumer pressure may create substantial transaction costs for a producer and a distributor by limiting the marketability of products that are objectionable to some consumers. Of course, the adoption of standards or production technologies that are consistent with consumer preferences also represents a transaction cost, but that cost is often outweighed by benefits in the form of greater marketability of the product. A distributor can easily pass that cost on to the producer by purchasing only from those suppliers who meet the standard and to the consumer in the form of a mark-up for offering a 'socially responsible' product. A producer, on the other hand, must weigh the cost of implementing the standard demanded by a particular market against the cost of finding an alternative market. The latter may be prohibitive if the product is bulky or otherwise difficult to transport, or if demand for the product depends to a great extent on diverse consumer preferences. In such situations adopting the standards demanded by consumers in the existing market niche may be a less costly option.

Changes in behaviour may also be prompted by an increase in the transaction costs of the 'old' way of doing business. In my view a market campaign against a firm significantly increases the transaction costs for the company of continuing business as usual, whereby it may prefer to change practices and take the FSC certification path. The more companies there are taking the certification path, the lower are the transaction costs for other companies to join. Therefore, a market campaign against a firm that increases the transaction costs of continuing the environmentally harmful practice helps to break down 'old' path dependence. Companies then prefer to change practices, adopt environmental policies, and get FSC certified. For the newcomers, FSC certification involves high transaction costs, but after adopting the practice these costs decrease. The greater the number of companies that become FSC certified, the better the new path dependence mechanism works.

I see an NGO-driven market campaign as not only as a form of governance, but also as a conflict that is deliberately initiated to contribute to desirable social changes. Social changes are expected in both transnational spaces and the spaces of place. In order to produce desirable social change all conflicts have to be managed by the stakeholders such that positive conflict potential increases and negative consequences decrease (Glasl 1999; Yasmi et al. 2006). The case analyzed below addresses the issues of conflict management in both transnational spaces and the spaces of place and explains stakeholder interaction in the relevant contexts.

Conceptual Model of a Cross-Border Market Campaign

The 'consumer campaign' to be analyzed here is composed of two conflicts: one in transnational spaces, the other in a concrete space of place (see Figure 9.1). The conflict started with the recognition that an actor (B-p) was involved in a practice

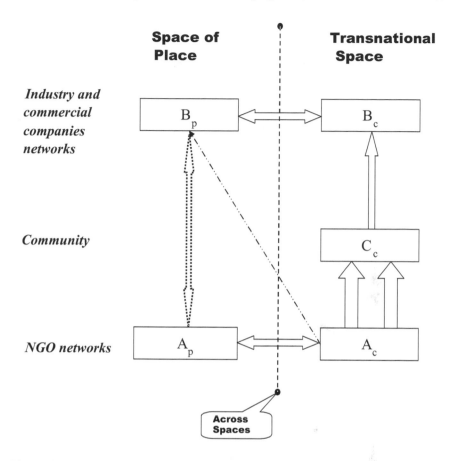

Figure 9.1   Relations Between Networks in a Market Campaign

that is detrimental to the environment. The NGOs in the space of place (actor A-p) monitored the activity of B-p and informed their counterparts in transnational spaces, actor A-c. The information about wrongdoing was passed across borders, to the consuming countries in particular. Actor A-c mobilized resources and organized a consumer campaign involving consumers (actor C-c), which targeted multinational corporation headquarters – actor B-c. A-c and C-c viewed the behaviour of B-c as detrimental to others. Therefore, they became involved in the conflict, which took place primarily in transnational spaces.

This changes the stakeholder interaction in transnational spaces and makes the transaction costs for doing business as usual too high for the actor B network. Actor B-c faces the consumer boycott, and prefers to force actor B-p in the space of place to change its practice. Again the conflict is shifted from a transnational space to a space of place. In the space of place many other stakeholders are involved in addition to actors B-p and A-p. However, A-p and B-p are essential for the conflict to happen. Other stakeholders include local, regional and/or national governments, local citizens and many others who make an effort to manage the conflict.

Further, the dynamics of the interaction of actors can change both in transnational spaces and spaces of places. In transnational spaces the effect of a consumer boycott of one company resonates and usually changes the behaviour of other companies. The 'old' path dependence is broken down and companies look for a new way of doing business. In order to avoid future boycotts, companies improve their social and ecological policies, develop programmes that demonstrate their corporate social responsibility, join associations of responsible producers, participate in eco-ratings, improve codes of conduct and certify their forest management practices. By doing so, they create a 'new' path dependence that is more environmentally friendly. The greater the number of companies joining the sustainable path, the lower the transaction costs they face. The market campaigns also impact the behaviour of investment bodies that use toolkits and other methods to maintain responsible investments. Responsible investments provide support for the newly established path and decrease the transaction costs of sustainable businesses. More companies join the sustainable business 'club', and the 'new' path dependence starts to work as a supporting institutional infrastructure.

In the space of place the stakeholder interaction usually changes completely and the outcomes are not always predictable. A company may change its practices and become more socially and environmentally oriented, or it might abandon the area after the consumer boycott, which results in job losses and reorganization of business networks in the area. Consumer boycotts can adversely impact local communities in forest settlements. However, when new business networks come into the space of place or old business actors change their practices, a new path dependence is established and local stakeholders ultimately benefit.

The Context of Transnational Spaces

The transnational space is structured by different kinds of stakeholder networks and their interests. The European Union (EU), for example, represents one of the nodes of governance in that it develops forest conservation policies for Europe and contributes financial support for projects that promote the designation of specially protected areas and sustainable forest management. In line with the above-mentioned policy, in the period 1999–2001 the European Union allocated money in the framework of the TACIS programme for creating four specially protected areas in Karelia, among them Kalevala National Park. This created a favourable context for successfully resolving the issue of preserving old-growth forests after the NGO-led consumer boycott against the logging companies.

From the start, NGOs in transnational spaces were also engaged in designing preservation policies that could affect the Russian environment. In the early 1990s, international environmental organizations, including Greenpeace, began discussing the preservation of old-growth forests on the border between Finland

and Karelia. Greenpeace led an effort to propose that the Karelian forests be included in the international UNESCO World Heritage List. The World Heritage Convention was adopted at the General Conference of the UN on November 16, 1972, and the Russian government is among its 168 signatories. The 'Greenbelt of Fennoscandia' would have been a joint nomination from Russia, Finland, and Norway, and would have included 20 forest massifs along 1000 km of the border region (Greenpeace 2001). In all, the area comprised 1.5 million hectares of forest, some of it virgin. This effort eventually failed, however.

NGO-led consumer campaigns unrelated to the particular campaign to save the Karelian forests have contributed to shaping transnational spaces. In 1990, market campaigns were one of the major strategies of NGO networks. Such campaigns were focused against companies in different sectors of the economy with well-recognized logos, such as Nike or MacDonald's, and were directed, for example, against sweat shops, the use of harmful pesticides in agriculture, companies using unsustainable packaging, and so on. In the forest sector, the best-known and significant campaigns were conducted by the Rainforest Action Network, Forest Ethic and Greenpeace to save the Amazon forests. NGO networks targeted not only corporations but also their investors. The most influential campaign was that directed against CITIGROUP and the Bank of New York.

In the early 1990s, environmental NGOs started monitoring the area of old-growth forests in Russia, including Karelia. NGO networks already had extensive experience of organizing similar campaigns. They learned to shape consumer preferences through international information campaigns and to raise international awareness of the importance of old-growth forests.

As a result of multiple campaigns, multinational corporations around the world understood the risk that NGO networks could cause to their business. Many companies have not only become sensitive to NGO grievances, but have started partnering with NGOs in sustainable forest management projects. This trend has led to partnerships such as WWF-Home Depot and WWF-IKEA. Such partnerships have facilitated the creation of environmentally sensitive niche markets and the promotion of forest certification systems.

Forest Stewardship Council (FSC) certification has become one of the most important regulatory forces in transnational spaces for stemming forest degradation. These voluntary, non-governmental global regulatory processes develop stakeholder-based standards and accredit independent auditors to assess forestry operations. The FSC approach creates standards based on ten principles of sustainable forestry that are designed to achieve environmentally appropriate, socially beneficial, and economically viable forestry. If the standards are met, the logging company receives the internationally recognized FSC certification and gains access to environmentally sound timber markets, such as those of Western Europe. The company can sell its products for higher prices but has certain obligations toward the local communities near its forests.

NGO networks thus had the institutions in place in transnational spaces to convert harmful companies (Cashore 2002) by discouraging them through market campaigns from damaging the environment and encouraging them to adopt sustainable forest management through FSC certification. The changes in transnational spaces also shaped the contexts in different localities around the globe, influencing governments, local NGOs and other stakeholders.

The Context of the Space of Place

a) National Context

Russia's forests cover 1.2 billion hectares, 69 per cent of the entire territory of the country. The Soviet regime attempted to expand the forest sector by launching extensive programmes for building pulp and paper mills. Most of the regional forest complexes of the Soviet Union were located in the territory of present-day Russia and the role of the mills was to satisfy the demand for forest products in other regions of the Soviet Union and many other socialist countries. Since the early 1990s, the Russian forest sector has been undergoing profound changes, mostly determined by new patterns of internationalization.

Since the 1990s, Russia's system of forest management has been in a state of constant restructuring. In 2000, President Putin closed the Federal Forest Service and gave its responsibility to the Ministry of Natural Resources. The ministry thus became responsible for both protecting and harvesting forests. The interactions between different divisions of the government were further complicated by shifting jurisdictions. In 2004, after Putin's re-election, restructuring of the ministries continued. In the last eleven years, nature protection units have survived five reorganizations. Not a single new nature reserve (*zapovednik*) has been designated.

The Forest Code was enacted in 1997, but already in 2002 there was a plan to change it completely. The new code was not enacted until January 2007 because of recurrent delays and disagreements between governmental agencies and civil society institutions. Constant reorganizations and restructuring caused inefficiency in governmental policies and created an institutional void in the governmental regulatory system. To some extent this opened a space for private actors and voluntary regulatory mechanisms to fill.

In the 1990s, after the opening of the borders of the former Soviet Union, an array of actors from transnational spaces, for example, multinational forest companies and international NGOs, entered the Russian political arena. Multinational corporations started buying Russian logging enterprises and building infrastructure to facilitate their entrance into Russia's economy. NGOs, specifically large transnational environmental organizations, also entered Russia and established active subsidiaries as quickly as commercial interests did. These organizations, bringing with them the flow of money, values, and ideas of nature protection from transnational spaces, officially entered Russia's political and economic spheres. Greenpeace came in 1992 and created a central office in

Moscow. Since then, it and other large environmental NGOs have tried to influence government policy, industry, and the environmental awareness of Russian citizens. The expansion of Western environmentalism into Russia since the early 1990s has brought with it ideas and concepts of nature conservation and techniques of natural resource exploitation developed by the science, industry, and the third sector of the US, Canada, and the European countries. In sum, a number of international stakeholders with different value systems, a different understanding of land use issues and opposite interests came into the Russian political arena and built networks with different kind of Russian actors.

b) Republic of Karelia

The Republic of Karelia is a heavily forested area of Northwest Russia. Since imperial times, forestry has been the republic's primary industry with an orientation towards export (Autio 2002). In 2006, forest production accounted for 55 per cent of the republic's total industrial production, with 60 per cent of all forest products being exported (Kozyreva 2006). Karelia's border with Finland is 700 km long and forms a large part of Russia's longest land border with Western Europe. While most of the country's forest resources were far to the east in Siberia and less accessible, Karelia offered huge tracts of virgin forest with proximity to the important timber markets of the West. In addition, Karelia's extensive waterways –over 27,000 rivers and 60,000 lakes – provided an effective means to transport logs and so helped orient the forest sector towards export (Gov. Karelia info 2007).

Under socialism, two zapovedniks were created in Karelia in the 1930s – 'Kivach' and Kandalaksha (Kantalahti). These encompassed 19,300 hectares of mostly forested land. In addition, Vodlozero and Paanajarvi National Parks and the region's many nature reserves contain some of the largest tracks of virgin forest in Europe. An area of 404,000 hectares in Vodlozero National Park was nominated for the UNESCO World Heritage List in 1998 but was not included.

Despite Karelia's economic importance, it remained fairly undeveloped up to 1917, with over 90 per cent of the population living in rural areas; today, 74.1 per cent live in cities and towns (Laine 2002). The Soviets had plans to rapidly industrialize much of the country, and Karelia's forests became an important resource for efforts to enter the world timber market. The forests were exploited rapidly and sold at low prices in Western Europe. Many researchers see this as a strategy of the Soviet government to acquire hard currency, which was much in need at the time (Autio 2002). As the country became increasingly centralized, economic planning for Karelia became disconnected from the regional level. Quotas were high and labour was often in short supply (Autio 2002). In addition, processing plants were overwhelmed and thus increasing amounts of unprocessed logs were exported. The central government owned all forests after 1930 and controlled all investment on the local level (Autio 2002a). Very few funds were put towards the development of forestry infrastructure in Karelia. In addition, the Soviet Union's economic approach to logging was based on the belief that increased harvesting means increased profits. This trend accelerated especially in the 1950s with increased large-scale industrial logging and clear-cuts.

This somewhat one-sided economic policy for the region had important impacts on the local level that persist until today. During the 1990s, Karelia experienced an economic downturn and forest production declined nearly 40 per cent (Autio 2002). This downturn somewhat reduced pressure on forests, but the government continued to promote the policy of 'more harvesting means more economy'. Thus, in the 1990s, faced with increasing economic difficulties, republic officials regarded increased exploitation of forests, especially of old-growth forests, as a necessary and urgent task.

Over 60 per cent of Karelia's virgin forests were harvested (Yanitsky 2000) in the course of the 20th century. The forest industry was especially interested in logging the old growth in the former military zone, where transnational NGO networks were interested in creating a new Kalevala National Park.

c) Local Context

The territory of the controversial Kalevala National Park represents a forested area that lies partly in the former Finnish border military zone and has only one small village inside it. The same families have lived in this territory for generations; they were relocated to a bigger town during socialism, but returned after the fall of the Soviet Union. Currently there is a small farm inside the park and small-scale tourism with Finland is developing. From the very beginning, local villagers were strongly against logging in the area, yet not interested in federal agencies coming to govern the park and build its infrastructure. The major local stakeholders lived in the villages of Voknavolok (Vuokkiniemi) and Kalevala, the small city of Kostomuksha or across the border in Finland, outside of the disputed area. There were no local NGOs participating in the debate on the Russian side of the border, but a number of Finnish civil society groups were involved.

d) The Finnish Side of the Border

There was also a conflict going on in Finland that centred on saving borderland forests. The commercial logging on the Finnish side of the border was much more intense as the Finnish military zone was much narrower than that of the former Soviet Union and thus only small spots of old-growth forests remains. Both international and Finnish NGO networks easily mobilized the local population in support of Kalevala National Park on the Finnish side of the border. As a result, the Finnish Kalevala National Park was formed, consisting of several small specially protected areas (Härkönen 2005, 202–213). People who were mobilized in support of the Finnish Kalevala Park became dedicated supporters of Kalevala National Park on the Russian side of the border; they adopted and understood in full the value of old-growth forests and participated in the actions in Russia organized by NGO networks.

The old-growth forests were not the only reason for the support of Finnish communities for the Russian Kalevala National Park. The Karelian borderlands are perceived by both Karelians and Finns as culturally close; the local people on both sides of the border share a common language, common traditions, common Karelian-Finnish epic runes and folklore. The *Kalevala*, the Finnish national epic was collected in the Finnish Karelian borderlands, especially in the area close to the park, and Elias Lönnrot (1802–1884) invented the title 'Kalevala' for the poems collected. The first edition of the Kalevala was published in 1835 and the second, with many more runes and songs, in 1835 (Lehtinen 2006, 175). Later on, the name Kalevala was given to the park.

Finnish people were thus eager to preserve old Karelian villages with their traditions and surrounding forests. In their perceptions, if forests were logged around Karelian villages, they would change and lose their ethnic spirit (interview, Finnish ethnographer, 2006). Finnish ethnographers, especially those studying folklore, collaborated with Russian scientists from Petrozavodsk and Kalevala and organized expeditions to Karelian villages to collect songs and describe traditions. Although they did not participate in direct actions, these people were strongly supportive of all kinds of preservation programmes in the area.

Finnish tourist firms developed special trips to Karelian villages, including areas inside Kalevala Park, in order to introduce tourists to Karelian culture and the simple lifestyle, which is still completely not modernized; Finnish people can live in Karelian villages as they might have lived in the last century (interview, Finnish tourist, 2006). The village inside Kalevala Park is a special attraction as it is home to a family whose ancestors were rune singers; their house is like a open-air museum in which people live and demonstrate old traditions, including animal sacrifice, that date from pagan times and were still common in remote Karelian villages in the nineteenth century. Both culture and pristine nature are advertised on such trips. Currently, a new tourist route called 'the Blue Road'[1] is being discussed that would link ethnic places in Sweden, Finland and Karelia. Kalevala National Park is on the route, between the other stops in the village of Voknavolok and Kalevala.

All these projects, although not directly related to the preservation of old-growth forests, constituted the context in the space of place for the success of a NGO-led market campaign organized in transnational space.

The Karelian Old-Growth Forest Market Campaign

The international NGO-led market campaign took place in the period 1995–2006 with the aim of saving the old-growth forests in the Republic of Karelia. This campaign was organized by Greenpeace and the Taiga Rescue Network in conjunction with Russian NGOs (the Forest Club), in particular the Centre for

---

1   The road is described as blue due to the many lakes along the way.

Biodiversity Conservation (CBC) and the Socio-Ecological Union (SEU). It also included the Nature Protection Corps, which began as student groups in Russian universities in Soviet times and involved students in combating illegal poaching and logging of Christmas trees. The campaign included numerous publications, videos, conferences, and protests. Using satellite images, the Forest Club inventoried and mapped virgin forests in the region. They investigated the timber sources for publishing houses in England, the Netherlands, and Germany, and requested that they boycott the logging of Karelia's old-growth forests. This culminated in 1996 in a series of publicized protests both in the forests of Karelia and at the pulp and paper mill of the large Finnish logging company Enso (Yanitsky 2000). This led to Enso's announcement of a moratorium on logging on three important sites in the disputed forests in Karelia. In 1997, several companies, both Finnish and Russian, joined the moratorium.

The Forest Club's message was manifold: they listed companies logging the old-growth forests, as well as the buyers in Europe that accepted wood from these companies. They implored the European public to boycott products made with Russia's old-growth wood. Companies logging old-growth forests in Karelia were breaking no laws or norms of the Russian Federation; however, NGOs were trying to enforce new informal global environmental rules and values beyond the control of any one state.

With the Russian government, the NGOs tried to initiate a process of creating a specially protected natural area in order to preserve the old growth forests. This last effort also witnessed the introduction of nature protection measures created in transnational spaces, including national parks, and UNESCO World Heritage Areas created by the UN.

After 11 years, the conflict was resolved with the victory of the environmental organizations. This was a transnational conflict based on the core belief of environmental organizations that the old-growth forests throughout the world needed to be preserved.

Latent Stage: The Foundation for Kalevala National Park

In 1995, in Kostomuksha Nature Reserve, environmental NGOs from 10 countries, including representatives from Canada, Sweden and Finland, held a meeting dedicated to the preservation of old-growth forests. At this meeting the Forest Club was formed, comprising representatives from the NGO Social Ecological Union, Greenpeace, Nature Protection Core and the Karelian student organization SPOK. The Forest Club started developing criteria for identifying old-growth forest and mapping the existing old-growth forest landscapes. It also started to monitor logging and cross-border trade operations in the area (interview, director of the Kostomuksha zapovednik, 2006). In 1996, together with the Russian NGOs Socio-Ecological Union and the Centre for Biodiversity Conservation, Greenpeace created maps of virgin forests along the Finnish border.

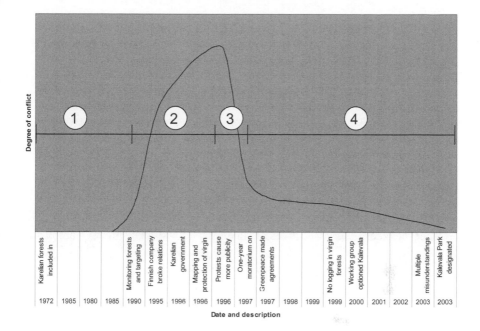

Figure 9.2   Old Growth Forest Conflict Stages

In 1997, in the framework of Russian-Finnish project on biodiversity conservation and sustainable forest management in Northwest Russia, money was allocated to the Kostomuksha Natural Reserve and the Karelian Scientific Centre for field research and for the designation of the borders of the future Kalevala Park. On the basis of a scientific report, the Karelian Committee for Natural Resources and state agencies in Russia and Finland had to make a decision on establishing Kalevala National Park (interview, Karelian Scientific Centre, 2006) (see Figure 9.2, marker 1).

Escalating the Market Campaign

In parallel with the effort to negotiate the creation of the national park, the Greenpeace network and Russian Forest Club started their market campaign. They informed companies in Great Britain and Germany about the actions of their suppliers from Sweden and Finland who were logging in Russia in the areas close to the future park. The maps of old-growth forests were presented to publishing companies and governments, including those of Karelia and Finland, as well as timber producers and consumers. Direct actions began in 1995 and the Finnish company Tehdaspuu was one of the first to break off its relations with a logging company working in the disputed 'green belt' in Karelia (see Figure 9.2, marker

2). However, in early 1996 the Karelian government supported continued logging in the area, citing economic interests (Yanitsky 2000). The next year, the chair of the government, V. Stepanov, issued the following statement:

> The government of the Republic of Karelia has to evaluate the actions of various ecological NGOs as interference in the internal affairs of Russia that is aimed at undermining both the Karelian and Russian economies and that violates the basis of the boundary policy of the Russian Federation (see Yanitsky 2000, 242).

Peak of the Conflict

A series of protests in August 1996 brought the area greater publicity. Members of Greenpeace blocked Finnish harvesters in the Kostomuksha district of Karelia and staged a protest a few days later in Finland at the pulp and paper mill of the Finnish company Enso (currently Stora-Enso) (Yanitsky 2000).

De-Escalating

This led to Enso's announcement of a moratorium, beginning January 1, 1997, on logging on three important plots in the disputed forests in Karelia. Enso promoted creation of a working group with NGOs and governmental representatives to discuss conservation issues and further inventory old-growth forests in Karelia (Lehtinen 2006, 187). In 1997, several companies joined the moratorium, including UPM-Kymmene, MoDo, Vapo, Kuhmo, Pölkky, and Ladenso (Vorobiov 1999). Greenpeace made agreements with a local forestry unit (*lespromkhoz*) stating that the latter would not lease the territories; however, it was already nearly impossible to find a willing lessee. Soon after this, a working group was formed to set guidelines for dealing with the virgin forests and for making future decisions on the issue. This group included environmental NGOs, government officials, and industry representatives. The Karelian government joined the group but again stated its opposition to the environmental NGOs when the chair accused them, in a letter to the deputy chair, of illegal acts aimed at undermining the Karelian economy (Yanitsky 2000) (see Figure 9.2, marker 3).

Period of Stagnation and Subsequent Resolution

In 1997, the Karelia Science Centre began preparing detailed justifications for the specially protected area of Kalevala. By 1999, there was no logging going on in the old-growth forests and the first round of justifications was finished. In the framework of the EU TACIS grant, a decision was made to establish the park on 95 thousand hectares. At the end of the project in 2001 the park had still not been designated, but the equipment for the park had been bought from grant money. A municipal agency, 'Kalevala Park', was established to manage the area during the transition period with the idea that when the park was set up the equipment would be forwarded to the federal administration of the park (interview, former director of municipal agency, 2006). In the framework of the Dutch-Russian programme Matra in 2002–2003, 1500 signatures were collected from Karelian citizens in support of the future national park and a set of recommendations was developed for park development.

Negotiations related to the park took place in parallel on different levels of governmental structures with deadlocks and delays due to conflicts between different kinds of state agencies. In 2001 an agreement on the park was negotiated on the level of municipalities; however, on the level of the Republic of Karelia the government agreed to a park with an area of 74.4 thousand hectares only (Yanitsky 2000). On August 6, 2002, the government finally issued a decree establishing the park and the requisite documents were submitted to the federal authorities. The period 2002–2006 was a stagnant stage in Karelia, as the decisions had been forwarded to Moscow.

In Moscow, the Kalevala Park documents were received just when state agencies were being restructured, and began their journey from cabinet to cabinet and from one agency to another. When the documents reached the stage of environmental impact assessments, it turned out that the procedure was not provided for in the state agency's budget, so Greenpeace contributed \$5000 towards making the assessment, as designation of the national park was essential for the 'happy end' of the market campaign (interview, local governmental representative, 2006). Such intervention seems a bit ridiculous – a radical environmental organization taking financial responsibility instead of a government. Finally, it was only in November 2006 that a positive decision was made on the federal level to set up Kalevala National Park and to allocate money for its infrastructure (see Figure 9.2, marker 4).

The maps of old-growth forests created by NGOs became a regulatory guideline for business operations. The Segezha pulp and paper mill rejected a lease for land adjacent to the future Kalevala National Park. Swedwood, a subsidiary of IKEA, signed an agreement with environmental organizations on a moratorium on logging old-growth forests in the area of the park. Following these informal agreements the total area proposed from the very beginning – 95,000 hectares – has now been preserved.

Stakeholders Involved in the Conflict

This conflict is about natural resources, yet all economic, social, cultural and natural considerations come into play. The conflict can be described as a multi-stakeholder conflict situated in both space of place and the transnational space. I frame the stakeholders of the Karelian Old-growth Forest conflict into two groups: stakeholders of the space of place, who have a stake and interest in the contested territory of Kalevala Park (both in Finland and Russia), and stakeholders in transnational spaces, who have a stake in what is happening globally with old-growth forests and for whom the particular territory of Kalevala Park is one example of a location where a global preservation policy has to be implemented.

Stakeholders' Interactions in the Space of Place

On the Level of the Karelian Republic: Karelian Government Agencies

The conflict involved many stakeholders. However, despite the fact that the conflict occurred mostly between NGOs and logging companies, after the moratorium on logging in the area the major arena of negotiation was left only to three major stakeholders – governmental agencies, scientists and NGOs – which were constantly blaming and shaming each other.

The Karelian government agencies and their supporters had several points of disagreement with NGOs:

a) Economic Concerns and Job Loss   The Karelian government stressed the financial importance of the republic's forest sector to the economy. One respondent said,

> I worry that if a specially protected area is created in Karelia, Russians will lose jobs. That's why in preserving old-growth forests we need to think about people. In Kalevala there are 150 people working in forestry. If these operations are discontinued, what will these people do? Who will create new jobs? (interview, Karelian governmental representative, 2002).

Another official said,

> Maybe they [environmental activists] are good and want to do good, but if we do as they want, there can be negative results, especially for our economy (interview, staff in Karelian Ministry, 2002).

They argue that Karelia already has a high percentage of officially preserved territories compared to other regions of Russia. My informants were supportive of protection in general and described it as positive, but they were reluctant, based on economic concerns, to increase the amount of protected area. One said,

> I say to the Finnish side, 'Why don't you create specially protected areas in other regions of Russia? And when they have the same percentage of protected area [as Karelia] then we can talk about new specially protected areas here' (interview, Karelian Ministry head 2002).

b) Financial Concerns   Government officials also complained that pressure from the international community was not supported enough by financial help for creating the specially protected areas. According to one respondent, the chair of Karelia's government wrote to the heads of various European countries requesting such funds (interview, head of Karelian Ministry of Natural Resources, 2002). He said, *'many didn't even reply. Some replied and said "we can't help you"'.* After a national park is designated and officially signed for, there are necessary expenditures, such as for tourism infrastructure. Concerning this, the respondent said,

> A huge sum is needed from the budget. In Karelia now, salaries are so low and there are unemployment problems. There's no oil and no gas, unlike in Komi. We are concerned . . . Who will invest all of this money?

The government claims to be in a bind, without enough money to fulfil the requests of the international community.

On the local level, the following statement by the town administration of Kostomuksha supports the same point:

> The town administration recognizes well the historical and cultural importance of this part of Karelia, but due to the critical situation in both Karelia and Russia, there are no funds for financing nature protection activities, or other important social expenditures.[2]

One of respondents from science also agreed, saying,

> We have two functioning parks and if it were not for international projects like TACIS, the parks would sit on the budget and nothing else. The TACIS programme significantly increased the salaries for the administration of protected areas, but such support is impossible to get permanently (interview, scientists, 2002).

c) Issue of Sovereignty   The government of the Karelian Republic tried to see this issue as a domestic problem, and felt that environmental activists from outside of Karelia and Russia had unjustly interfered in the republic's issues. For the Russian government, the overwhelming interest of European NGOs and European governments in Karelia's forests was not readily explicable. In my interviews with government officials, I heard various assessments: NGOs are saboteurs trying to undermine Russian forestry for the benefit of Scandinavian competitors, or NGOs are exaggerating the urgency of protecting Russia's virgin forests and biodiversity, both of which currently abound. Whatever the discourse was, a prevalent accusation made by all government officials was that Europe had logged nearly all of its own

---

2   Letter from the mayor of Kostomuksha to the Taiga Rescue Network, 1996.

old-growth forests and was thus creating a double standard by forcing Russia to preserve those which remain.

An independent scientist agreed with this last sentiment, saying,

> [Kalevala Park] will have significance for all of Europe. They've cut everything on their territories and we will preserve on our territory. In their own country they use forests like orchards, with trees in rows, and on our territory they want to use it for recreation so that they can see wild nature and have fun (interview, scientist, 2002).

d) Ability to Compromise   The government accused environmental activists of not compromising. One government official said,

> We always go from compromises to understandings. But our opponents also need to compromise and understand . . . It doesn't always happen though, and we are blocked and not understood. We are not understood by Greens (Interview, state committee 2002).

In addition, many officials were suspicious of the NGOs' goals. One respondent said,

> Sometimes we think that in the bottom of their hearts they have their own interests… Greens sent us maps coloured in green and they absolutely don't want to explain why these forests are virgin. They say just 'from satellite pictures'. We can conclude that they are acting in somebody's interest (interview, state consultant 2002).

About the maps he also said,

> Whenever an enterprise here begins to work in international markets, its territory is always coloured by Greens.

e) Lack of Science Behind the NGO Position   Government officials claim that much of the NGO position in this conflict was not backed up by science, or that it was based on incorrect science.

f) Value of Old-Growth Forest   One official argued with the science of preservation in general,

> They [NGOs] say don't touch the forests, they belong to the planet. But if the forest becomes older and older then tomorrow they will take all our oxygen and we will not get it back. This is what happens with old-growth forests (interview, Karelia Ministry of Natural Resources, 2002).

Another official argued for the benefits of forest management over total preservation,

> In Europe there are no virgin forests, but I don't think this is bad. You must be very careful when dealing with old-growth forests. If they are not taken care of, they'll become problematic areas. Forests that are not taken care of – when you don't cut sick plants –become problems for forestry. When forests are old it is like in life when old people are alone. Sure I don't mind if old people exist, but it is bad when they are alone. You need young people to help them. If young people don't care about the old, they will die. A forest is like a living body and if it is not revitalized it will become older and diseases and pests will spread and this is bad (interview, state consultant, 2002).

On the Level of the Karelian Republic: The NGO Network

The NGO network that had a stake in setting up Kalevala National Park operates partly in transnational spaces and partly in the space of place. The Forest Club and Greenpeace participated in local direct action, worked with the Russian government on the federal level, and took the lead in the Internet consumer campaign. Only a few NGOs in Karelia, SPOK being the most active, interacted intensely with governments on the level of the republic. The major disagreements were both conceptual, especially regarding the preservation of old-growth forests, and procedural, related to the interaction itself.

a) Old-Growth Forests have a Planetary Value   The Forest Club, led by Greenpeace, has tried to establish the concept of a virgin forest both in the legislation of the Russian Federation and in the awareness of industry and the public. The goal is to convince stakeholders in the forest that virgin forests have a value in the West and therefore must be preserved in Russia. The attempt to import this idea into Russian industry and government has not proceeded smoothly, because Russia, unlike Western Europe, contains vast stands of virgin forest and old-growth forest landscapes are not that unique.

b) Soviet Mentality of the Russian Government   Environmental NGOs see the mentality of the Karelian government as harmful to the population and the republic's economy. Our respondents said that government officials still have the mentality from Soviet forestry that 'more logging equals more money'. The NGOs criticize this viewpoint and the lack of sustainable forest management in general. One informant said,

> They [Karelia's government] have never made an assessment of how many forests are left; they live under the illusion that there are still plenty and that's why our forest resources have become weaker and weaker (Interview with NGO CBC representative, 2002).

c) Lack of Effective Forest Policy Both independent scientists and environmentalists alike see a problem with the republic's forestry strategy, and so feel that even if the old-growth forests are logged and not preserved, they will not bring economic benefits to the republic.

d) Government is Corrupt and Represents Industry Interests Some NGO representatives see close connections between the government and forest producers as a reason why the government wants to allow old-growth forests to be logged (interview with CBC coordinator, 2006). In addition, one respondent further criticized the head of the republic and Karelian politics in general as corrupt:

> Decisions there are made by a very narrow circle of people close to power, and they make them without any logic ... the Karelian government is a museum of socialism (interview, Greenpeace 2002).

e) Government is not Doing its Job NGOs in Russia frequently must do technical work in place of the government. A SPOK activist said,

> In the West . . . you spread knowledge about the problem and immediately the public begins to participate. The authorities get kicked and then they understand the problem well and begin to do something . . . there is no action from Russian power structures. . . To achieve something here you need first to make a big noise, and then secondly you need just to do everything yourself, instead of the government. And then you will achieve results (interview, NGO SPOK).

In both the government and NGO narratives, it seems that there were many gaps in communication between the two parties. The impossibility mentioned by NGOs of arranging a meeting with high-level government officials is one illustration. All of our NGO informants explained this by saying that the government is simply unwilling to cooperate with them. In the years of dispute over this issue, representatives of the environmental position were unable to meet, even once, with the governor of Karelia, Leonid Katanandov. One respondent said,

> They tried very much to arrange meetings. And when the meeting was set up, they [the Karelian government] send only assistant governors who were available at that moment. This came to no agreement. They came to the meetings and were saying 'yes yes' and shaking their heads. Or they would show up and say that they cannot have the discussion today. Sometimes they were just silent and said nothing. Sometimes they said simply 'we don't support the issue'. But a meeting was never arranged with Katanandov himself, whose status allows him to say either yes or no,. This is just not understandable (interview, NGO SPOK, 2002).

Similarly, environmental activists failed to meet with the head of the republic's Ministry of Natural Resources, another government official with many responsibilities linked to nature protection. About him one of my informants said:

> It is absolutely impossible to reach him or arrange a meeting. He does not talk with representatives of NGOs. It is his principle. He doesn't even say hello. His vice chair is polite and he always says that he cannot solve the problem because he has no power or because he does not have his own personal opinion. So it is impossible to have a normal conversation with them (interview, Greenpeace 2002).

Many environmental activists at the late stage of negotiations saw miscommunications with the Karelian government as a big hindrance to the creation of Kalevala National Park. The local administration in two regions, scientific organizations, the local forest management unit (*leskhoz*), and industry had agreed to the creation of the park; however, Governor Katanandov did not sign the proposal. This would have been the final step in Karelia towards the park's creation. Attempts to contact the government and find out what was causing the hold-up met with delayed and vague responses. One of our informants said:

> For each letter, according to the rules, they [the Karelian government] have a whole month to respond, so they wait the whole month and then send a letter . . . After a month we get responses like, 'yes, we wrote to Katanandov and he will analyze the situation and reply to the government and to you and explain what exactly is going on' . . . but nobody does anything. Everybody just has correspondence and nothing is done (interview, SPOK 2002).

In the late stages, Karelian government officials worked together with representatives of science and the environmental movement in a working group on this issue; however, our NGO respondents felt that this cooperation was not genuine. One of the informants explained governmental delays in decision making by saying:

> The decision is not made because their ideology is to measure things in cubic metres. It is real big money and what exactly they can get from the park is unclear (interview, Karelia Scientific Centre 2002).

Local Stakeholders of the Space of Place

On the local level the key stakeholders on the Russian side of the border were local administration, the local forest management unit, the local community, the Kostomuksha natural reserve, scientists from the Karelian Scientific Centre of the

Russian Academy of Sciences who came periodically to do research in the area and logging companies themselves.

There were no local NGOs in the area. The input to the space of place by NGOs from the outside, especially Greenpeace, was strategic. They organized short, direct actions. For example,

> they brought a bus with journalists to the area, attached themselves to the tractor and put up posters. The journalists took pictures and everybody left before the police came – very fast (interview with local state authority 2006).

It was only when the central TV stations showed the report that people found out that the direct action had taken place, that local stakeholders became engaged in the issues and that intense conversations and negotiations between interested parties took place. Following the scandal, the Forest Club representatives again came from their Moscow office and held professional negotiations with local stakeholders.

> They were very flexible, when someone behaved aggressively with them; they never responded aggressively; they carefully explained their position on the issue (interview, local administration, 2006).

The local administration did not express any open support for the direct actions taken by NGOs, but they generally argued in favour of creating Kalevala National Park and participated in the negotiation process intensively. After the TACIS grant money ran out, they helped establish the municipal organization 'Kalevala Park' and for several years paid a salary to its director. They also argued that the former military infrastructure that was left on the Finnish border should be part of the park in the future (interview, Kostomuksha mayor 2006). The people's deputies council in Kostomuksha also supported the establishment of Kalevala National Park and new candidates stated their support for the park during their election campaign.

Local people were supportive of the park from the outset. Most local people do not perceive logging operations as possible employment, as modern technologies do not need too many people. For the local people, the development of tourism, as well as using the forests for recreation as well as gathering mushrooms and berries, is essential. The village inside the proposed park is very small and all villagers were strictly against logging, although they were concerned about hotels that might be built and destroy the spirit of the village; they were also worried by the federal infrastructure that might be built in the park. One local farmer was interested in developing family tourism with Finland, revitalizing old Karelian traditions and running a family business on his own without any interventions (interview, local community representative 2006).

The local state forestry unit (leskhoz) in Kostomuksha felt frustrated about the moratorium on logging in the territory of Kalevala Park. It argued that if a geographical area continues to be in its jurisdiction, it should be leased to the logging company and logged. They did not mind the creation of the national park or ceding the land to federal jurisdiction, but they were against having the responsibility of managing the area during the stagnation of the long-term moratorium (interview with director of local forestry unit 2006). Their major frustration was related to the fact that forest companies signed informal agreements with NGOs on moratoriums in the areas of old-growth forest on the territories of their leases using Greenpeace's maps as common law (interview, forest unit employee 2006). They blamed both NGOs and companies for neglecting governmental agencies in such decisions. They viewed NGO direct actions sceptically and believed that the NGOs were interested in conducting such actions not to save the forests, but because they were receiving money from the West (Interview, director of local forestry unit 2006).

From the very beginning, the Kostomuksha Nature Reserve supported both the preservation of old-growth forests and the creation of new specially protected areas. They provided support for NGOs, for example, by hosting the NGO international conference that took place in 1995. At this conference the goal of preserving old-growth forests was articulated and the Forest Club was formed. The Kostomuksha Nature Reserve also participated in the research conducted in the Kalevala National Park in order to provide a scientific basis for the creation of the park. The Karelian Scientific Centre of the Russian Academy of Sciences followed up the research done by the Kostomuksha Natural Reserve and supported the establishment of the park in the Kalevala area.

In the 1990s, logging companies were strongly against establishing new specially protected areas in the border regions in Karelia. These areas with massive old-growth forests became accessible for logging operations after the fall of the Soviet Union; previously the land was part of the military zone along the border. This land became desirable for lease by Russian, Finnish and Swedish firms, but they were faced with a consumer boycott. This case saw virtually no interaction between NGOs and industry beyond the consumer campaign of the 1990s. All international companies working in the disputed old-growth forests of Karelia abandoned their leased territories. No company, Russian or foreign, would apply to log these forests, and so NGOs had no further business with industrial stakeholders on the local level, except those who had leased the land around the 'moratorium' zone.

In 2005–2006, the logging companies that had leased the land around the Kalevala Park area completely changed their attitude toward the old-growth forests, taking the path of FSC certification with compulsory preservation of high-value conservation forests, especially old growth. The big companies, such as Swedwood Karelia (a subsidiary of IKEA) and the Segezha pulp and paper mill made an agreement with the Forest Club on a moratorium on logging of old-growth forests and on buying wood from old-growth forests (interview, SPOK 2006). These informal agreements between companies and NGOs made the total area removed from commercial use, including Kalevala National Park, 95,000 ha (the officially designated park on the governmental level is 74,500 ha). Therefore, the total amount of land now set aside is that originally proposed by environmental NGOs in the early 1990s, although the area was diminished by governmental decree.

Interaction of Stakeholders in Transnational Spaces

In transnational spaces large transnational environmental NGOs with their transboundary networks have tried to stop the Finnish and Swedish logging industry from destroying the old-growth forests and to influence government policy in Russia by forcing the government to establish a new national park. The case demonstrates that their interaction with multinational corporations proved to be more successful than that with Russian governmental agencies. The Taiga Rescue Network (with its major office in Sweden) and Greenpeace were the key actors in transnational spaces that brought the message from the Russian Forest Club to the stakeholders in transnational spaces. They were the key players in organizing the consumer boycott against logging of old-growth forests in Karelia. By encouraging European buyers to boycott products from Russian old-growth forests, Greenpeace effectively eliminated the threat of logging by forcing business actors to change practices. As in most Greenpeace-led campaigns, the media were essential for shaping the issue in transnational spaces. Through the media Greenpeace used the ecological sensitivity and environmental conscience of European buyers, appealing to their values in order to mobilize consumers to participate in a boycott. Values based on discourses in transnational space addressing biodiversity conservation and old-growth forests preservation have been used as an instrument for purposefully escalating the conflict with the aim of social change.

In the beginning, the conflict was barely recognized by business actors in transnational spaces. The logging companies were not aware of the problems involved in their activities. The Finnish and Swedish companies operated legally and their subsidiaries came into the country by invitation of the Karelian government and officially leased the territory. They never expected the conflict to happen. However, after media reports and direct actions, multinational corporations with subsidiaries in Russia reacted relatively fast by establishing the moratorium in the disputed area. Many of the business partners of Russian logging companies in early 2002 asked their Russian suppliers to get FSC certified. This request had a multiplier effect and was directed not only toward companies in Karelia, but many others in Northwest Russia.

The European Union, with its TACIS grant programme, was another important stakeholder of transnational spaces. In the framework of the Convention on Biodiversity Conservation, the European Union established the priorities for the TACIS programme, which channelled money into countries of Eastern Europe and Central Asia. Money for the establishment of protected natural areas along Karelia's border with Finland came from the TACIS programme in 1999–2001.

TACIS gave the Karelian government 3.5 million dollars to establish two new national parks, one of them Kalevala, and to build a tourist and nature protection infrastructure for the two already in existence. There were many tensions and miscommunications between the actors in transnational spaces and actors from the space of place, especially representatives of the Karelian government, in the framework of this transaction. According to our informant, Karelia has acquired a negative image in the international community from this controversial issue and its handling of it. He said,

> In Europe Karelia is a scandal region where no one can agree on anything. They only have conflicts there and scandals (interview SPOK 2002).

The case demonstrates the lack of communication between the TACIS programme and Russian governmental officials and scientists. The lack of communication affected negotiations in Russia and by the end of the grant period none of the planned parks had been designated; only related research had been carried out.

Conclusion and Implications

In the market campaign, NGOs operated bottom up from the space of place to transnational spaces. Accordingly, the Karelia old-growth forest case can be divided into two phases by time: the conflict that resulted from the consumer campaign in transnational spaces took place earlier then the conflict in the space of place. In the escalating stage, the space of place was used only for short direct actions with the aim of getting images delivered to transnational spaces. The acute phase at the peak of the conflict was quite short (1995–1997) compared to the long stagnant conflictive stage in Karelia afterwards. When access to the natural resource was restricted by a moratorium on logging in the Kalevala area, the conflict shifted overwhelmingly to the space of place and continued there for almost a decade (1997–2006).

The information campaign in the transnational space was essential for transforming business practices. In the course of the market campaign, NGOs in transnational spaces used both normative and coercive pressures to foster change in corporate behaviour and to institutionalize the new practice. Without the belief that the ancient forests are of planetary value, it would be not possible to mobilize consumers to boycott the logging companies. Simultaneously, NGOs applied coercive pressures by naming, shaming and blaming the companies that were responsible for destructive activities in the old-growth forest in Karelia, and introduced new values while appealing to consumers at large. Before the NGOs started the campaign there was no international awareness about old-growth forests. This awareness was accelerated when organizations like Greenpeace started their information campaign. By doing so they institutionalized new consumer preferences based on ecological standards and facilitated the future change in companies' path dependence. It is important to notice that the information campaign took relatively few resources compared to its significant outcomes. Only four NGO representatives in Moscow and one in Karelia were needed to frame and deliver the message to the transnational NGO networks who conducted mobilization in transnational spaces.

Both normative appeals to consumers and coercive pressures on the logging companies resulted in increased transaction costs for the companies in doing business as usual, prompting them to change their practices. Currently all European companies that are involved in wood trade in Russia are aware that there is a possible threat to their image if they buy wood harvested from old-growth forests. They request transparency from Russian companies on the origin of the wood. Many companies in Northwest Russia took the forest certification path in order to gain legitimacy in the European market. In 2006 Russia took the second place worldwide after Canada in the amount of land where forest management is FSC certified.

Maps of the old-growth forests created by the coalition of NGOs were used as a tool for land use decisions and preventive measures. They became an informal law for logging companies, and compliance with this law became a kind of licence to operate in Russia. Companies that leased areas containing old-growth forests made informal agreements with NGOs and established voluntary moratoriums in these areas even though Russian governmental agencies were pushing them to pursue logging (interview, Swedwood, 2006). My case study has demonstrated that the institutional change occurred first in the transnational spaces and later in the space of place. As a result of the campaign the logging companies in Russia were forced to make changes in their operations, as their business network requested them to do so.

As a result, the market campaign to save Karelian old-growth forests had a significant overall impact on the space of place and on forest management in Russia. In the 2000s, logging companies went through profound restructuring. Almost all small logging enterprises become subsidiaries of large multinational holding companies that have a recognized image in the international arena. These large holding companies recognized the risks related to logging of old-growth forests and contributed to the institutionalization of new practices.

For the Republic of Karelia, however, the effect was twofold. On the one hand, on the positive side, there was no negative impact on local communities. This conflict took place in areas which were not heavily populated by local people. Those who were employed by logging companies continued with their jobs in places close by, but where forests were not old growth. On the other hand, the market campaign resulted in long-lasting misunderstandings between regional stakeholders, especially between NGOs and governmental agencies, as there were not any effective procedures for conflict management. This case helped to shed light on the difficulties and complications of bringing global practices of sustainable forest management to Russia. The situation was not recognized by governmental agencies in the beginning as a threat to doing business as usual. NGO influences were not known in Russia, where, in a young and newly established democracy, policy making still maintained a traditionally strong top-down approach. Because of institutional instability and turbulence and endless reforms it was not possible to designate the park until the year 2006. During the period 1997–2006 it was not possible to carry on commercial logging, either, or to build tourist infrastructure; in sum, it was economically unfeasible to keep the territory unused. The local state forest management unit was forced to do maintenance work, such as fire protection, without any economic rewards. Even those financial flows that were coming from the transnational spaces in the form of TACIS grants were not used properly. Instead of implementing the new park in the space of place, the funds were used only to do related scientific research and even the equipment that had been bought for the park was kept frozen. However, these financial flows alleviated the tensions that occurred between the stakeholders in the space of place, getting local governmental agencies to favour the park. What was peculiar in this particular case was that Kalevala Park represented a contested issue not only for Russian stakeholders, but for Finnish stakeholders as well. The participation of Finnish stakeholders in both debates and protests contributed significantly to the successful outcome.

This case demonstrates the asymmetry in the institutional development of stakeholders in transnational spaces and spaces of place. When NGOs operate bottom up and direct their grievances to the actors in transnational spaces, stakeholders in transnational space react in the way NGOs expect them to react. When the transaction costs of doing business as usual are greater than making changes, companies implement changes. Change in business practices then move forward and are implemented in the space of place, in Russia. However, the market campaign in transnational spaces does not target national governments. Governmental agencies continue to operate according to the old path dependence and, even worse, in the situation of institutional turbulence, become a significant barrier to implementing innovative institutional changes.

References

Autio, S. (2002), *Forests and the ecological dimensions of industrialization of the Soviet Union in the 1930s* (BASEES).

Autio, S. (2002), 'Soviet Karelian Forests in the Planned Economy of the Soviet Union, 1928–37', pp. 70–90 in Laine, A. and Ylikangas, M. (eds).

Arthur, B. W. (1994), *Increasing Returns and Path Dependence in the Economy* (Ann Arbor: The University of Michigan Press).

Bernstein, S. (2001), *The Compromise of Liberal Environmentalism* (New York: Columbia University Press).

Cashore, B. (2002), 'Legitimacy and the privatization of environmental governance: how non state market-driven (NSMD) governance systems gain rule making authority', *Governance Journal*, N.15, 4 October.

Cashore, B., Auld, G. and Newson, D. (2004), *Governing Through Markets: Forest Certification and the Emergence of Non-state Authority* (New Haven: Yale University Press).

Castells, M. (1996), *The Rise of the Network Society* (Cambridge, MA: Blackwell).

DiMaggio, P. and Powel W. (1983), 'The Iron Cage Revisited: Institutional Isomorphism and Collective Rationality in Organizational Fields', *American Sociological Review*, 48(2): 147–160.

http://www.gov.karelia.ru/gov/Info/eco_geo3_e.html, updated March 2007.

Greenpeace Russia (2001) *World Natural Heritage Areas in Russia*.

Härkönen, K. (2005), 'First steps of the Kalevala Park' in Tervonen and Härkönen (eds), pp. 202–213.

Kozyreva, G. (2006), *Problemyi formirovanya sotsialnykh institutov ustoichivogo lesoupravlenya* (Petrozavodsk: Karelsky nauchnyi tsentr, RAN).

Laine, A. (2002), 'Continuity and Change in 20th Century Russia', pp. 22–28 in Laine, A. and Ylikangas, M. (eds).

Laine, A. and Ylikangas, M. (eds) (2002), *Rise and Fall of Soviet Karelia* (Helsinki: Kikimora Publications) B 24.

Lehtinen, A. (2006), *Postcolonialism, Multitude, and the Politics of Nature* (Lanham: University Press of America).

Lehtinen, A.A., Donner-Amnell, J. and Saether, B. (eds) (2004), *Politics of Forests: Northern Forest-industry Regimes in the Age of Globalization* (Aldershot: Ashgate).

Meidinger, E.E., (2003a), 'Forest certification as a global civil society regulatory institution', in Meidinger, Elliot and Oeasten (eds), pp. 265–289.

Meidinger, E.E., (2003b), 'Forest certification as environmental law making by global civil society', in Meidinger, Elliot and Oeasten (eds), pp. 293–329.

Meidinger, E., Elliot, Ch. and Oeasten, G. (eds), Social and Political Dimensions of Forest Certification (Freiburg: Germany Verlag).

Tervonen, P. and Härkönen, K. (eds) (2005), Kalevala Parks, Life on the Border and Pristine Nature (Kajaani: Kainuun Sanomat).

Tysiachniouk, M. and Reisman J. (2004), 'Co-managing the taiga: Russian forests and the challenge of international environmentalism', in Lehtinen et al. (eds) pp. 157–175.

Urry, John (2000), *Sociology Beyond Societies* (London: Routledge).

Vorobiov, Dmitriy (1999), *Transboundary Movement Toward the Saving of the Karelian Forests, in Towards a Sustainable Future: Environmental Activism in Russia and the United States* (St. Petersburg: Institute of Chemistry of St. Petersburg State University).

Yanitsky, O.N. (2000), *Russian Greens in a Risk Society* (Helsinki: Kikimora Publications).

Chapter 10

Transformation of Nature Management in Pomorie: Fishing Villages on the Onega Peninsula of the White Sea

Antonina A. Kulyasova and Ivan P. Kulyasov[1]

Introduction

In this chapter we analyze the transformations that have taken place in the nature management institutions of Pomor fishing villages on the Onega Peninsula of the White Sea.[2] We focus on changes in traditional nature management practices, primarily with regard to fishing, as this is the main source of employment in these areas and brings about the development of a certain type of culture. We look at social institutions in the Pomor villages from a historical viewpoint, as well as analysing their current situation.

In order to analyze the institutional forms of organization with regard to fishing in Pomor villages, we apply the theory of path dependence, first formulated in the mid-1980s in the work of Douglass North (North 1990). This theory was then developed in the work of such authors as Klaus Nielson, Bob Jessop and Jerzy Hausner (Hausner, Jessop and Nielsen 1995), and also Oliver Williamson (Williamson 2000).

Through their work, the researchers put forward two approaches: on the one hand, social institutions characteristic of a particular society tend constantly to replicate themselves over the course of the society's development, thereby hindering both their own transformation and the emergence of new institutions,

1   We owe thanks to Hugh Fraser, Nadezhda Mukhina, Lois Kapila, Barbara Seppi, Eleanor Bindman and John Riedl for the translation of this chapter and Barbara Seppi and Edward Saunders for the proof reading of the translation.

2   The historical territory of *Pomorie* comprises territories of Russian North in the Onega, Dvina, Sukhona, Mezen, Pechora, Kama and Vyatka river basins, which have a specific culture. In our days a smaller territory is more often regarded as Pomorie – the coastal territory of the White Sea and Barents Sea in the Arkhangel'sk Region, Karelia Republic and Murmansk Region. The Pomor cultural-ethnic group is identified on the basis of the territory they live in and traditional economic practices such as deep-sea and costal fishing, seal hunting, river and lake fishing and forest hunting. Fishing still remains the main economic activity of the Pomor people; most of the money they earn comes from fishing.

and thus movement always proceeds along the same path. On the other hand, the 'path dependence' approach presupposes that, through the activity of various social agents, motion along this path can be corrected and overcome. In order for this to occur, social agents must create a new, strong direction of movement, which will bring about the creation of new institutions. In this way, the current path is rejected (path-shaping) and a new path is taken up.

In our chapter we will be applying both of these approaches, but focusing on the first, looking at how historical forms of organization in society and nature management which developed during the Soviet and pre-Soviet periods have been retained through the period of reform. We will be focusing above all on transformations in the fishing kolkhoz, a social institution that has been the most stable organizational form for collective management for more than 70 years. By analysing the current situation and the problems relating to nature management in Pomor fishing villages, we allow the current survival strategies of these communities to be distinguished. In addition, because there is a strong link between traditional fishing and the Pomor identity, we will also be looking at contemporary means for seeking ethnic identity.

The chapter is based on field material collected over the course of three expeditions in November 2005, February 2006 and November 2006, on the Onega Peninsula of the White Sea and in the cities of Onega and Arkhangel'sk. We would like to thank all respondents who allowed us to interview them and who provided us with materials. We would like to express particular thanks to the chairs of the fishing *kolkhozy* 'Belomor', '40 Let Oktyabrya' and 'Imeni Lenina', and the heads of the non-governmental organizations – the chair of 'Pomorskoe Vozrozhdenie' and the head of 'Aetas' – for their help and contributions. For purposes of the research, high-quality sociological methods were used. During the expeditions, biographical and semi-structured interviews were conducted, materials were collected, and travel notes kept. Respondents included members of fishing kolkhozy (the chairs, workers and pensioners) and other local people, representatives of local and regional administrative bodies, members of local government, village public sector workers, private businesspeople, scientists, and leaders of Pomor non-governmental organizations.

Respondents were selected on the basis of the social structure of the village, of which the fishing kolkhoz is the basic institution. Most people in the villages – including both workers and pensioners – are members of this organization. The remainder are public sector workers, working at the school, club, library or medical station, but also private businesspeople, berry pickers, hunters and anglers, unemployed people, and regular seasonal residents (whose main place of work or residence is elsewhere). Here we should point out that the fishing kolkhoz is not only the main social institution in a Pomor village, structuring both its economic and social life, but that it also determines the main forms and practices of nature management. Of course, nature management is not limited to the kolkhoz, but in this contribution this form will be the focus of our analyses.

As mentioned above, we will be analysing transformations in communities in Pomor fishing villages with the help of the theory of path dependence or, as it is referred to by Russian sociologists, 'track theory' (Auzan, Radaev, Nureev, Naishul et al. 2006). Looking at the Pomor village as a community whose economy is based mainly on the traditional use of natural resources, we see a limited, fairly stable set of year-round and seasonal practices. This stability can also be seen in the social institutions that help to organize the life of the community.

If we look at these social institutions from a historical viewpoint, we can observe a gradual change in the elements that make them up, but also a continuity. Often, new forms of social institutions include elements taken from the past. For example, the first fishing *kolkhozy* absorbed characteristics from pre- and post-revolutionary fishing workers' cooperative associations. It could be said that the life of people in the Pomor village develops 'along a familiar track,' or 'according to path dependence'. Yet at the same time, global modernization processes introduce new technologies into old forms of nature management (Korotaev 1998), bringing about changes in the 'path'.

But this takes place according to a process not so much of substitution, as of addition. The transformation of old social institutions and the emergence of new ones in Russia over the past decade has had virtually no effect on institutional forms in Pomor villages. The last radical change in social institutions in Pomor villages took place when the kolkhozy were first formed. Thus we can speak of the inertial potential of communities in Pomor fishing villages as being characteristic of a traditional society.

At the same time, the current phase of transformation of Pomor communities reveals the possibility of a new fundamental change. The state is radically changing the rules of play in the primary area of nature management – fishing – but also in forestry and other activities, changing the laws with the aim of strengthening large businesses operating in the fishing and forestry industries. This directly affects life in the Pomor communities, and weakens their developing foundations (Titova 2006).

In contemporary scientific literature, much has been written about the current state of traditional nature management, both in Russia and elsewhere. Russian writers generally describe traditional nature management as conducted by indigenous peoples in the Russian North, Siberia and the Far East, analysing the relationship between traditional nature management and traditional lifestyles (Mangataeva 2000). They analyze the legal aspects of the rights of indigenous peoples to the territory on which they practice traditional nature management, as well as approaches to protecting these rights, taking as an example the conflicts that arise between the market interests of large extracting companies, and the goal of protecting the homelands, culture and identity of indigenous peoples (Novikova 2003).

Much has been written with regard to preserving the cultural-natural landscape and the sacred places of indigenous peoples, and preserving the peoples themselves (Vedenin and Kuleshova, 2001). One aspect that is dealt with is the relationship between the preservation of culture and community, in other words between the preservation of traditional nature management practices, and the ethnic identity of the indigenous people (Arkhangel'sk Region Territorial Community of the Pomor Indigenous Minority, 'Pomorianaya Storona' 2004). Articles and books on contemporary nature management have also analyzed state policy in this field, for example regarding problems facing the fishing industry in Russia (Titova 2006), and the management of natural resources in the Barents Region (Averkiev et al. 2004).

Traditional Pomor nature management has generally been examined from a historical viewpoint. Cod fishing and other types of fishing in the Barents Sea have been analyzed (Krysanov 2002), as have the state of seasonal industries and their significance for life in villages on all shores of the White Sea (Popov and Davydov 1999). The remaining forms of nature management in Pomorie – inshore, river and lake fishing, sealing, hunting, reindeer herding, salt-making, agriculture and forestry – have also been analyzed, generally from a historical point of view.

This chapter is unique in applying path dependence methods to analyze problems relating to both the transformation of the current state of traditional nature management, and to the fishing industry. We believe that the nature management conducted by the communities in Pomor villages, which is currently divided between fishing kolkhozy and the private sector, is both a part of the fishing industry, and a traditional form of Pomor nature management.

It is important to note that there is great uncertainty surrounding the status of Pomorie and of the Pomors, who are not included in the state's official list of indigenous minorities, and to whom legislation grants none of the rights of native minorities to conduct traditional nature management. Thus Pomorie does not officially exist as a territory.

However, in the last full Russian population census in 2002, some 6,500 people cited Pomor as their ethnic group (most these people live in Arkhangel'sk Region). Most of the population did not know that the Pomors have at last been added to the list of nationalities in Russia; in other words, the Pomors did not know that they would be officially counted if they cited their ethnicity.

There are still debates surrounding the acceptance of the Pomors as an ethnic group, among both scientists and politicians. A variety of constructs of the Pomor identity exist, including dictionaries of the Pomor language (Moseev 2005). There are non-governmental organizations in Arkhangel'sk Region fighting for the Pomor population to be granted the right to ethnic self-determination – NGOs 'National Kultural Avtonomia Pomorov', 'Pomorskoe Vozrozhdenie', 'Obshchina Pomorov,' etc. Although all of the required documents have been submitted, the Pomors have still not been accepted as an indigenous minority at the federal level. There are different viewpoint on the question of Pomor ethnic identity – '*who are Pomors?*' – among scientists. Some of them, such as Doctor Bulatov, head of Pomor State University, see Pomors as a separate ethnic group based on Finno-Ugric and Slavic (Bulatov 1999a). Some others, according to Doctor Bersham, a well-known Soviet ethnographer who conducted extensive research on the Pomors, identify them as ethnic subgroup of North Russians (Bershtam 1978). Most specialists in Russian traditional culture and the history of the Russian North identify Pomors as belonging to the North Russians (Vlasova 1995). But independently of how they identify the Pomors all researchers emphasize that the Pomor population have a specific lifestyle, including traditional nature management practices.

History

The territory of Pomorie came into being over several centuries. Officially, *Pomorie* had formed by the sixteenth century, comprising territories in the Dvina, Sukhona, Mezen, Pechora, Kama and Vyatka river basins (Bulatov 1999b). By that time Pomorie had developed into a separate, unified region characterized by a

> common territory, access to the sea, common features of economic activity in its administrative districts and towns, particular character traits, the specific mental and spiritual makeup of the Pomor people and a distinctively northern culture. The northern form of the Russian language also developed, providing the basis for contemporary local tongues and dialects (Bulatov 1999b).

The question of the identity of the Pomors (coast-dwellers) and its relation to their traditional usage of natural resources can now be considered. As the Pomor ethnic group, distinct since the sixteenth century, has never been recognized as a nationality (ethnic group), attitudes towards it have always been to some extent ambivalent. Moreover, the ambivalent approach to their ethnic identity is shared by the Pomors themselves and their neighbour peoples. Non-Pomors consider the inhabitants of this region to be Pomors and Russians simultaneously, but they still note the peculiarities of their character, culture and use of natural resources.

This unique ethnic group (the Pomors) is identified on the basis of the territory they live in. They are people who live in a particular territory, say, in a coastal village, and have particular occupations, such as deep-sea and costal fishing and seal hunting, among others. The Pomors do not necessarily consider themselves to be Russians. They associate their identity with the territory they live in and their economic practices. But they still continue to regard themselves as Pomors even if they move to another place or give up fishing.

As already mentioned, Pomorie is a sizeable territory, but our discussion here will be limited to the coast of the White Sea. In centuries past, the main occupations of its inhabitants were fishing and seal hunting. Fishing still remains the main economic activity of the Pomor people; most of the money they earn is from fishing.

In the nineteenth century, the Pomors were engaged in several types of fishing. The main activity was cod fishing in Barents Sea near the Murman coat, in which a significant number of Pomors from all parts of the White Sea took part. Cod fishing began in May and ended in late autumn. Other occupations included seal hunting in spring, as well as costal and river fishing. These occupations are regarded as traditional because they were carried out using traditional fishing equipment. There were small rowing boats for sea fishing. Moreover, the traditional character of fishing is confirmed by the Pomor concept of fishing as akin to harvesting, something which can only be carried out at a particular time of the year. This concept of fishing is reflected in the saying 'the sea is our field' (Maximov 1984). The traditional nature of this trade is also confirmed by the use of sacred magic rituals during fishing, various incantations and ceremonies (Korotaev 1998). Fishing existed in this form until the early 1930s, when collective farms, the kolkhozy, began to be established in Russia.

Russia's transition to a market economy and competition from Norwegian fishers initiated a transformation of traditional methods of nature management. However, there were no crucial changes at that time, and the Pomor traditions were still upheld. The one thing that helped conserve the traditional way of fishing was the Pomors' resistance to the modernization of fishing boats and equipment. This trend emerged among the Pomor people at the turn of the nineteenth century. It is interesting that Russian Pomors communicated very actively with the Norwegian Pomor people, who were the driving force behind this modernization.

> Russian Pomors watched the Norwegians. They neither rejected everything, nor did they adopt everything. For instance, Russian Pomors rejected trawling, because they considered it destructive (Korotaev 1998: 49).

The traditional ways of nature management by the Pomors, including fishing, was based on the Pomor traditional culture – a dialogue between human being and nature, which was viewed as an intelligent power.

The physical requirements of deep-sea and costal fishing determined how it was organized. Since fishing tackle was sizable, even in small boats several men were needed to handle it. For this reason, fishing had to be carried out collectively, by a group of people or a team. In other words, the collective nature of Pomor fishing determined its organizational form. Thus, in the middle of the nineteenth century, the practice of '*pokrut*', or employment, emerged. This was when wealthy Pomors, who owned boats and fishing equipment, hired a team of poorer men, who were called 'pokrutchiks' (Krysanov 2002).

At the beginning of the twentieth century, the Russian state promoted cooperatives that were independent of wealthy boat owners, providing loans to them to enable them build boats and buy fishing equipment. Pomors also organized cooperatives for coastal ice-fishing and established common settlements for fishers. Seal hunting was also carried out collectively. The fishers acquired all the necessary knowledge in the traditional way – from previous generations (Korotaev 1998).

It should be pointed out that, in addition to fishing, the Pomor village residents were also engaged in farming and various forms of forestry, which involved whole families and even clans. Social interactions were regulated by a traditional village community called 'mir' (Vlasova 1995: 8). Thus, the Pomors were simultaneously fishers, seal hunters, peasants, hunters, sealer, gatherers and wood-cutters. But their main activity, fishing, still prevailed. This traditional way of life in a Pomor village was maintained for centuries (Bershtam 1978). The most significant changes that took place in the Pomor villages in the twentieth century were establishment of fishing kolkhozy in the late 1920s and 1930s. During that period, the Russian village was transformed. The Soviet regime tried to eradicate traditional ways of life and agricultural practices in Russian villages and make it exclusively utilitarian from an economic point of view. As a result, the social and economic life of the Pomor villages deviated from its traditional forms for the first time in this period as a result of the policy pursued by the Soviet state. During that period, the main social and economic institution of the Pomor village, the 'mir', was forcibly destroyed and replaced by collective farms (Korotaev 1998).

As regards the history of Pomor collective fishing farms, in the early thirties, collective farms became the main social institution. In this context, social and economic activities, including traditional nature management, were developing in the Pomor villages. Collective farms were established and traditional agricultural methods were transformed, a development which we will now examine.

The Transformation of Collective Fishing and Other Fishing Practices in the Soviet Period: The Collective Farms of the Onega Peninsula in the Arkhangel'sk Region

Practically all of the fishing kolkhozy on the Onega Peninsula appeared in 1930. Originally, almost the entire adult population joined collective farms. One of our respondents described a kolkhoz in the following way,

> It is neither a private enterprise, nor is it a state one, but a productive cooperative. People combine their material portions and their labour. And with these they can raise capital (Head of a fishing kolkhoz, m., b. 1946).

The main distinguishing factor with regard to the creation of fishing kolkhozy in coastal villages was the presence of the tradition of collective fishing, which had been in existence for a considerable length of time, during both the pre-revolutionary and post-revolutionary period. '*We had artels here, you see; it is difficult to work alone at sea, it's hard on the family*' (Hunter and fisher, former teacher in Purnema village, man, b. 1960). The actual process of organizing the collective farm was, as in other places, extremely difficult. Compulsory collectivization of the fishers' boats and vessels, livestock and equipment took place; some were dispossessed ('raskulacheny') and sent into exile. However, the ready-made social structures, namely the fishing artels, made the transition to the kolkhoz collective system of farming considerably easier.

Although the fishing kolkhozy were created as fisheries, their members were also forced to do a considerable amount of agricultural work. One of our respondents observed, '*At that time, we had gosplans [state plans from State Planning Committees] and party tasks*' (Chair of the collective fishery, man, b. 1946). They began to cultivate new crops and breed animals, which had previously not traditionally been bred. Alongside the traditionally cultivated cereals and legumes, members of the collective farm started to cultivate vegetables, including potatoes, while those involved in livestock began to keep poultry and swine, as well as the traditional horse-breeding and sheep and cattle rearing. The agricultural work was carried out by teams of workers. Often adolescents were among those who laboured in the teams.

Traditional methods of fishing in the collective fishery were gradually transformed. By the end of the 1930s and beginning of the 1940s sea fishing was already dominated by the use of small vessels – boats with mechanical engines. In addition to this, sailing and rowing vessels (lod'i, ran'shiny, shnyaki, yaly, karbasy) were still being used, but significantly fewer were being produced. On several motorboats, a trawler net was already being used for catching fish. This was considerably more lucrative than traditional seasonal fishing with lines,[3] sweep-nets, riuzhi,[4] and nets. These boats, nevertheless, had limited movement, and like those with sails were unable to venture further than 40 miles from the shore even in good weather. Fishing with them, therefore, continued to be limited. All the fish which were caught were processed by the crew by hand before being passed on. The amount of fish caught was comparatively small, and thus this form of fishing did not undermine the fish stocks (Korotaev 1998).

It is necessary to note that fishing using small vessels and motorboats is fully compatible with the historical view of traditional exploitation of nature. To a certain extent, this could be considered comparable to the transition from traditional hunting methods to the use of firearms. Furthermore, the rigging used on the boats was in principle the same as that used on sailing boats. Such fishing practices can therefore be considered traditional, to the extent that they were not alien to the coastal village community. Previously, members of the collective farm participated in the fishing, and only the captain and the mechanical engineer were engaged as trained professionals, having completed special training elsewhere. The remaining members of the team learned the skills of seafaring and fishing from the previous generation, and studied them within the community.

In addition, one segment of the population of the communal farm practised sea fishing while another practiced coastal and stream fishing, both when working and in their free time. The traditional sacred relationship to fishing surprisingly quickly gave way to a utilitarian approach. Prayers and incantations faded from everyday usage, but did not disappear completely, remaining at the back of people's minds.

---

3   Lines – fishing tackle comprised of a rope, several hundred metres long, and bait, usually small fish, on hooks.

4   Riuzha – a variety of net, used in shore (ice) fishing.

Such customs were still adhered to by the older generation. Often, when they found themselves in extreme situations, both the older and younger generations resorted to these customs (Filatov 1994).

Besides all this, the members of the collective farm were busy with agriculture. It is necessary to note that agriculture was first and foremost directed towards ensuring that their own needs were met; the remainder of the produce went to the state, although it was always unprofitable due to the lack of satisfactory transport links. One of our respondents characterized their agricultural production in the following way,

> There was agriculture, just as there is today. It's been preserved to a small extent. But all this was only supplementary due to the profit that we received from our fleet of boats. Agriculture was fundamentally unprofitable; we didn't make any profit from it. Both bringing things here and transporting them away is very expensive. You just end up over-paying. It was to feed yourself as well. We handed over some to the state, of course. We had to fulfil the [state] plan (Head of a fishing kolkhoz, b. 1958).

Work on the collective farm was extremely hard, especially during the Second World War and its aftermath, as the majority of the male population had perished. One of our respondents recalled this period,

> In spring, we ploughed collective fields for 30 days. I ploughed using a couple of horses. And then sawed all the firewood for the collective farm and for ourselves, and then there was haymaking in remote areas. We were cutting, rowing, making stacks. Autumn and summer pass in the same way, we dug up the potatoes, transported them, ploughed the fields. Then in winter, in the fishing season, we caught fish in very remote places (Pensioner, former worker at the collective fishery, b. 1918).

During this period, fishing was continued by the women, under the direction of the elderly men. The women not only occupied themselves with the shore fishing, but also went out on the boats. One of our respondents remembers it in this way,

> I was also at sea for a year, and spent the winter in Murmansk. On the 3 November, such a strong storm swept in from Pertominsk. It was impossible to eat or drink. Then, another year, I only arrived on the 14 November, and spent a whole year on the boat. We set off from the collective farm to fish for herring with a sense of duty. Even if you didn't want to, you went (Pensioner, former worker at the collective fishery b. 1922).

The geographical location of the fishing villages determined their daily routine in many respects. The majority of them were isolated from the wider world, did not possess reliable transport links, and were out of range of all electrical supplies and radio and telephone links.

Accordingly, the provision of transport links depended upon the season. In summer, access was from the sea, and in winter over land. It was expensive to use aeroplanes and helicopters, therefore helicopters were deployed only in cases of emergency. Settlements had their own mini power stations. Then, in the Soviet period, coastal villages had an image of being isolated communities; there was an increase in communication between villages, but significantly less with the rest of the mainland, the regions and the administrative centres of the districts.

Notwithstanding the isolation of fishing settlements, the state system of handing over and processing fish was organized in such a way that it worked well for the collective farms. Besides this, the kolkhozy had their own association, – a fisheries cooperative – within the framework of which they resolved all issues pertaining to fishing, the sale of fish, loans for repairs and the acquisition of vessels. Moreover, this administrative structure and system stayed in place until the end of the 1990s.

One distinctive feature of the fishing kolkhozy throughout the Soviet and post-Soviet period was their profitability. The practice of fishing gave the upper hand to the collective fishing farm over the purely agricultural kolkhozy, which in the 1950s and 1960s were converted into *sovkhoz* farms (state collective farms) en masse; in other words, they became state enterprises. This phenomenon was brought about due to the state's new investment policies, directed first and foremost towards the support of unprofitable state farms in areas where farming was unreliable. One of our respondents described it in the following way:

> Soviet agriculture appeared, and the state began to help. Capital was invested, and not only directed towards production but also towards the social sphere, with the aim of developing the sociocultural daily sphere. The government made significant investments in this. The make-up of the feed changed a lot then. A lot of sugar beet, sunflower residues and feed-combinations were brought from the south; in other words there was a massive injection of resources into the agricultural industry (Head of a fishing kolkhoz, b. 1946).

This period took its toll on the collective fisheries. Several collective fisheries were also forced to abandon fishing and become state farms. At the same time, collective farms were enlarged. From 1930 to 1957, the territory of the collective farm overlapped with that of the village and its fields and woodland. In 1957, however, a government decree enlarged the collective farms to the extent that they absorbed three to four villages. They have remained in this form until the present day (Interview with the head of a fishing kolkhoz, b. 1958).

The process of enlarging collective farms and the change in the organizational structure of several collective farms (kolkhoz) into state collective farms (sovkhoz), took place in the context of the technical modernization of the fishing industry. In spite of their considerable profitability, additional means were needed for the modernization of the collective fisheries, namely loans. Loans were only given through the fishing kolkhoz union, or 'Rybakkolkhozsoyuz' and with the agreement of the party leader of the Arkhangel'sk district, and therefore not all collective farms successfully acquired credit (Interview with pensioner, former head of a fishing kolkhoz, f., b. 1922).

Towards the beginning of the 1970s, lines of communication were laid through the territory of several collective fisheries in the Onega region. These railway lines and highways meant that their agricultural activities became relatively more profitable. This was one of the reasons for the transformation of the fishing kolkhozy into agricultural sovkhov state farms. One could therefore see in these villages the substitution of one fundamental social institution for another. The new institution was essentially different from the previous one.

It is important to note that all six kolkhozy on the Onega Peninsula remained collective fisheries. In other words they preserved their fundamental social institute, that of a kolkhoz. In this same period, these fishing kolkhozy undertook a qualitative change in the structure and organization of the exploitation of natural resources. From the end of the 1950s to the 1970s, the fishing kolkhozy gained possession of large fishing trawlers. Through their union, the fishing kolkhozy hired professionals from elsewhere to work on the boats. At the beginning, those workers in the collective fisheries who had undertaken specialized training worked on these boats, but by the mid-1990s, on the whole, only hired workers operated them, although they did not become members of the fishing kolkhoz. One of respondents described it in the following manner:

> The crews of both the large trawlers and medium-size trawlers assembled in Arkhangel'sk and Murmansk. In other words, the colleges prepared entire workforces for the fishing fleets there. Our rural folk also went there, but as a rule they started as sailors then studied to become skilled technologists and mechanical engineers; they then became residents of the town (Head of a fishing kolkhoz, m., b. 1946).

The modernization in the methods used to catch fish led to a general change in the practice of fishing. Ocean fish were caught, and as a result a rift appeared between catches and traditional places and seasons. Kolkhoz boats were now out fishing throughout the year, working not only in the White Sea and the Barents Sea, but also venturing into the Atlantic Ocean. In addition, the quantity of fish caught increased, partly due to improved technology, and fishing was now practised on an industrial scale. It is easy to understand that such a view of the exploitation of natural resources was subject to a quota, determined by international agreement. In relation to this, one of our respondents noted: '*From 1975, we were limited by quotas, each year this quota grew smaller*' (Head of a fishing kolkhoz, m., b. 1958).

A rather paradoxical situation arose, as members of the fishing kolkhoz were transformed from workers and participants in a cooperative into proprietors of boats, quotas and fish. The kolkhoz, as a single collective proprietor was making a profit, and paid for moorage and boat repairs at ports, paid the wages of the vessel's crew, and for the team of fishers, while also organizing the sale of fish. The fishing kolkhoz spent the remainder of the profit on provisions for members of the fishing kolkhoz and other inhabitants of the coastal villages. Furthermore, the kolkhoz collective farms and all agriculture were maintained solely by the profit from the fish caught using trawlers. The farms produced dairy and meat products and cultivated vegetables and grain. Beyond this, the profit made from sea fish was also used to maintain the bureau of the kolkhoz and other kolkhoz housing, paid the wages of all members of the kolkhoz, was used to buy a tractor and other machinery and paid for the upkeep of the local diesel power station, which continuously required additions and deliveries of fuel oil (Interview with the Head of a fishing kolkhoz, f., b. 1967).

The 1970s and 1980s were a period of economical and social development, both for the fishing kolkhozy and the coastal villages, which were provided for by the profit from the fishing industry and fishing resources. Simultaneously, fishing with trawlers also provided the collective fishery with considerable revenues, and the politics of taxation of the Soviet government permitted the kolkhoz to use the revenue for the development of their infrastructure and social sphere.

> The trawler worked, agricultural production worked, there was a lot of livestock. All the same, the collective farming produced a profit and little by little started to build up a material base. We built one farm, then another; we built a garage, a boiler room, and then a nursery. The Soviet government began to help with the children's nursery. Before that, we hadn't had any help from them (Head of a fishing kolkhoz, m., b. 1946).

At the same time, the kolkhoz showed itself to be a self-governing social institution. Decisions pertaining to construction projects in accordance with the rules of the kolkhoz were taken at communal meetings of members of the kolkhoz.

> It was always like this, there were always discussions at any time. It was a democratic procedure. It always was, is and will be. I might be expected to say, as the chair, that this made things difficult, but it didn't. Discussing these questions meant that we discovered the weak and the strong points. In other words, I always ask the question, Are we making the right decision? It's good when there are suggestions. And it's bad when there aren't, when you come across a passive response to this or that question (Head of a fishing kolkhoz, m., b. 1946).

It is impossible to reconcile fishing practices using trawlers with traditional exploitation of natural resources; trawling is a purely commercial, industrial practice. However, it was precisely this practice which has provided security for the traditional life in coastal villages for the previous 30 years or so. It is possible to discuss, at this point, how within the fundamental social institution, new practices appeared which did not lead to a change in the usual path ('path dependency') or departure from the usual route ('track') of social practices and provided for the reproduction of institutions. We have seen how, over the course of several decades, a complete change occurred in fishing technology, in its participants and in the quantity of fish caught. It should be noted that, in spite of the modernization of engineering and technology, the same institutional form was preserved – the kolkhoz and the social practices of the exploitation of natural resources within the village.

The modernization of sea fishing did not result in a change in the fundamental forms of activities or the social institutions of the coastal villages. The functioning of the kolkhoz remained on its usual 'track' and new forms of fishing and agriculture were built upon this. But the onward development of coastal fishing villages with the collective farm as the fundamental socioeconomic institution was impeded by the reforms of the 1990s and substantially changed at the beginning of the twenty first century. We will analyze this period in the next section of this chapter.

Contemporary Life Strategies of Fishing Kolkhozes and Coast-Dwelling Communities

We highlight a few life strategies of coast-dwelling communities that were adopted in the reform period (1990s and beginning of the twenty-first century) to ensure reproduction of the community. The main strategy was the retention of the kolkhozy as the main social-economic institution. And within this institution we further distinguish strategies that helped the community to survive and to continue its development. Among the strategies adopted by the community was the creation of an indigenous minority and community council.

As noted above, most agricultural kolkhozy existed until the 1950s and 1960s and were transformed into sovkhozy, that is, into government businesses, many of which operated at a loss. In the fishing kolkhozy this did not happen. On the one hand, their basic activity, fishing, was yielding high profits, so these kolkhozy were considered to be rich. On the other, their structure and activity corresponded to their historically established collective form of nature management. For these reasons the kolkhozy survived the first wave of economic reforms, whereas in the middle of the 1990s all sovkhozy were handed over to different shareholders and turned into non- governmental stock companies, many of which went bankrupt and were liquidated.

In our view, the conservation of the kolkhozy as economic organizations and as the Pomor community's form of nature management tells about the peculiarities of the Pomor character. Ethnographers (Maksimov 1984, Terebikhin 1993) and respondents said it to be characterized by the absence of a business spirit in its contemporary understanding as the maximization of profits and outperforming all competitors. As a consequence, the reforms of the 1990s had no immediate impact on the socioeconomic conditions of the Pomor villages.

This gave the fishers of the kolkhozy a readiness to practise a real fishing economy and maintain themselves to a large extent. As a matter of fact, the fishers in the kolkhozes were self-financing throughout the Soviet period. They had only their own resources to invest in new technology and infrastructure and were able to provide themselves with basic vegetables, meat, dairy and fish products. This way the economical reforms could not prevent the Pomor villages from developing in their other, customary way ('track'). The main social establishment of the village could be well preserved thanks to the organizational forms that were necessary to adapt the reforms.

However, the reforms forced the kolkhozy to search for new partners (to look for a direct exchange with foreign countries), create new organizational forms (connect with each other, set aside part of the kolkhoz production for the use of the kolkhoz population), return to traditional agricultural technology (switch from tractors and cars to horses) and start making additional income by exploiting the forests; this meant that they had to start manufacturing, an unusual development for the most prosperous period of the fishing kolkhozy.

Despite the active search for living strategies, the fishing kolkhozy as enterprises operated at a loss for several years. Therefore they could not immediately restore their agricultural production under the new taxation system. One of the interviewees criticizes the policy of the government:

> As Chubays says, no matter how much money we send to this village, it all disappears. Therefore I could only tell him: don't send anything, just don't oppress us with taxes (Head of fishing kolkhoz, m., b. 1946).

In 1998 high taxes, inflation, liquidation of the whole administrative structure and of the system of the fishing kolkhozes' union led to substantial difficulties, for example in the activity of the kolkhozy or in their interaction. Some kolkhozy kept working independently; five kolkhozy united their fleets, establishing an 'association founded on trust, without the constitution of a juridical identity' and founding a management structure – a fleet management service consisting of hired specialists. This way, a new organizational structure emerged, one founded on free will and one in which the kolkhozy themselves decided on the extent to which they wanted to follow the fishing kolkhoz union. One institutional form appeared in the place of another, taking in previous patterns and, now as before, proposing a preset course.

Voluntary merger of the kolkhozy, as well as the inclination to work with foreign partners, proved to be an effective strategy. This allowed the kolkhozy to pay off debts within two years and to begin renewing the fleet structure in the year 2000 – improving the management and the fish-catching by hiring specialists and leasing new vessels that made it possible not only to fish for usual types of fish, but also to fish on the ocean. A respondent said:

> We were able not only to fish for cod and Peter's fish, but we also started fishing for herring, mackerel and perch. And thus we started buying boats, new models. We borrowed money from the government. The interest rates were high; even

today they are not low. But we developed in terms of structure; we started cooperating with the Norwegians, meaning they became our partners. Selling fish has always been more convenient abroad than in Russia (Head of fishing kolkhoz, m., b. 1946).

Life in the fishing kolkhozy was eventually put into order; wages could be paid on time and the agricultural equipment was modernized. However, the government was preparing new difficulties. In 2004 it started new 'nature preservation' reforms that reduced the fishing quotas for the fishing kolkhozy while increasing fees for biological resources. For the Barents Sea, the biological resources were redistributed between Russia and Norway by the bilateral Russian-Norwegian Commission. Until 2004, the laws were such that Russia distributed these resources among different subjects of the Federation, every one of which received shares according to their concrete activity. This distribution of the fishing resources is called the distribution of the fish catch quotas. (For more about fishing quotas, see Chapter 5.)

After 2004, fish became a federal resource. Therefore the setting of fees for using this resource and the distribution of the quotas among enterprises came under the direct control of federal institutions. The quota for each enterprise was determined according to a 'historical principle', or, as our respondents think, according to 'pseudo-historical' principles, since quotas were formulated according to the data available for the past three years. This way the main fishing enterprises, large industrial undertakings that had been actively participating in the quota auctions, received a large share of the Russian quotas. This is the manner in which quotas were distributed until 2008.

Under these circumstances, the majority of the fishing kolkhozy received small quotas (permission to catch 100 to 1,000 tons of three different types of fish, each year), since many of them were not able to take part in the auctions and could not buy quotas. In essence, the kolkhozy will receive their small share until 2008, and even afterwards the Russian government plans to continue this same form of distribution. One respondent explained:

> Whoever has received little will still receive little for the next five years. This is our hopeless situation – we are locked in a circle. We cannot participate in the auctions, we have no means for that, and we have no priority rights as an indigenous minority (Head of fishing kolkhoz, m., b. 1967).

To further explain this situation, we will show how the quotas were redistributed using the example of one kolkhoz. From 1998 to 2003 the kolkhoz received quotas of 900 tons of cod each year and as it owned a trawler, it could basically fish all year round. In 2004, the quotas went as low as 240 tons of cod and 130 tons of haddock, that is, 370 tons of sea and ocean fish per year. In addition to this, the government introduced restrictions. If a business or the fishing kolkhoz owned less than 620 tons of biological resources in their fleet, it was not allowed to go to sea. This forced the fishing kolkhozes to unite their quotas with other kolkhozy and companies, to reduce their fleets or to rent them out. As a result, the kolkhozy have been making less profit than if they were to use their quotas independently. This has had effects on the agricultural and social conditions of the Pomor villages.

In the course of these developments, large fishing businesses and different holdings asked the kolkhozy to sell their quotas or change their organizational form so that they could become part of the holdings. The kolkhozy did not take this step, understanding that this would make them lose their independence. The kolkhozy had not gone bankrupt and were not liquidated under extremely hard economical conditions. The kolkhoz collectives, supported by the entire village community, made a conscious choice and stayed on the 'beaten track', going the 'former route'. In contrast, the government's policy was to force the fishing kolkhozy to abandon their beaten track.

Not only does the governmental policy not support the fishing kolkhozy, but it also intensifies their competition with new private companies. This is how one of the respondents evaluates them:

> Once this was not a private traders' business. There were no private traders here. There were only large companies and the fishing kolkhozes. That's all. But then, in the 1990s, when this ended, private traders appeared everywhere, here and in the Far East. And sure, they also needed to make profits. They most likely also had friends in the government. They were able to lobby for their interests in the Duma. And that's it. They decided to pool everything, and again everything was redistributed at our expense (Head of a fishing kolkhoz, m., b. 1958).

It thus becomes clear that it is hard for collective fish farms to compete with major companies and their factories in a modern market economy. From an economic point of view, a collective fish farm, which is burdened with 'attendant industry' in the form of agriculture supporting community infrastructure, is not the most effective type of enterprise in the field of fishing, nor is it supposed to be. A kolkhoz has a different purpose – the welfare of its members. In this case one must understand the dual nature of the collective farm: on the one hand it is a business, on the other it is an association of local residents, a cooperative and a community which secures their everyday life, and even, to a certain degree, a form of autonomy. As a result, kolkhozy are not trying to make as much money as possible; they want to make enough to secure the lives of people in the community. When one considers that kolkhozy are virtually the only enterprises in coastal villages and that life in these settlements is entirely dependent upon them, one can understand the enormous role they play in preserving the coastal population. The issue is not just the preservation of local traditions and culture, but the preservation of these communities in general, since at present no real economic alternative for their development has been put forward.

Market reforms and the policy implemented by the state thus created a situation in which fishing kolkhozy found themselves in a difficult position. This prompted them to adopt a vital strategy which promised gains but kept to the beaten track as before. By uniting in an association based on trust, five fishing kolkhozy became fairly major players in the region's fishing industry. The kolkhoz association went to an auction and bought up more quotas so that there would be no less than 1,000 tons on board, bearing in mind that their boats could catch 1,500–2,000 tons per year. This would allow them to achieve the main aim of the activities of the fishing kolkhozy: to direct the profits from fishing towards preserving the lives of the population of their villages. One of the respondents spoke about this directly:

> In the village farming is not profitable. We sell electricity to the population for 80 kopeks, but its cost is 20 roubles. The state guarantees that we will receive the difference, but while everything is fine on paper, as soon as you get down to business the state says 'sorry, we don't have the money in the budget right now, the money will come later (Head of a fishing kolkhoz, b. 1958).

Another respondent added:

> Here we pay for all the training and the journeys to treatment centres, there and back. Not to mention that for pensioners every celebration and funeral is at our expense; it happens thanks to the kolkhoz. And local maintenance of roads, bridges, and berths as well (Head of a fishing kolkhoz, m., b. 1967).

In order to guarantee their competitiveness, the collective fish farms started to draw up another new and vital strategy which consisted of working directly with foreign partners. This strategy was relatively new and as before it incorporated existing experience. Russian Pomors had lengthy experience of pre-revolutionary cooperation with Norwegian coastal dwellers. From the 1930s to the 1990s, however, direct cooperation was not put into practice. During the reform period of the 1990s the conclusion of direct agreements with Norwegian, Swedish and other foreign enterprises and direct yields in foreign markets not only provided additional profits, but also developed direct sociocultural links. The strategy of working with foreign partners gave collective farms the opportunity not just to receive prepayment (of up to 80 per cent) and sell fish in Norway, Denmark, England and Portugal at a higher price than in Russia, but also to receive credit for the modernization of their fleet at more favourable rates than were possible in Russia. A respondent explained:

> While here we would get credit at a savings bank at 19 per cent annual interest, they would give us credit at 9 per cent interest. It was a noticeable difference. And they helped us to obtain, or rather, lease these modern boats. We started to build .processing plants there, and as a result we gradually renovated our fleet (Head of a fishing kolkhoz, m., b. 1946).

Yet another strategy which has enabled the viable development of fishing kolkhozy is the additional production which has come about in the past few years

– seaweed harvesting, catching seals, fish cultivation, and timber exploitation (timber harvesting and lumbering). Let us look at timber exploitation in more detail. Timber exploitation for both personal needs and for sale existed in coastal villages even in the pre-revolutionary and Soviet pre-collective farm period. It was traditional. Timber exploitation for the sake of the collective farm was carried out during the entire period of the farms' existence, but it became an independent form of production only in the 1990s in a few kolkhozy. A respondent told us:

> Around 1994–97 we cut down a forested area here and transported it to Murmansk with our fleet. At that time there were roughly 30 boats in the Union of Collective Fish Farms system. We took the wood and planks there and brought back fish. You could take a truck with 10–12 lots of timber and bring back 10–12 tons of cod, haddock, flounder or halibut. And of course the whole village and the town were pleased that the collective farm workers were finally bringing back fish again and feeding everyone with it. It was great (Head of a fishing kolkhoz, m., b. 1946).

As we have already mentioned, virtually all kolkhozy cultivate woodland and use the timber for their needs. Another respondent explained the other uses of timber by collective farms:

> We have to provide wood in the first place for our schools, clubs, libraries, hospitals, clinics, kindergartens, and administrative buildings. Then the collective farm members and pensioners who have the rights receive free wood from us; in other words we fully supply them with wood for the year. We do this in accordance with a decision made by the general assembly. And in this way we have protected our population throughout these 14 years of reform (Head of a fishing kolkhoz, m., b. 1946).

The opportunity to export lumber and lumber products all year round is very significant. But the volume of timber harvested by the collective farms is 10–20 times less than that produced by timber companies in the Arkhangel'sk Region. Even those kolkhozy which sell lumber and products do not produce more than 10,000 cubic metres per year, part of which goes for the renovation of collective farm homes and buildings and part of which goes on sale in processed form and as logs, with roughly a quarter of this wood for the coastal village's population and social facilities. Sometimes the wood is sold to businesses in the town of Onega. One collective farm which is situated next to a highway managed to sell its prepared timber in Onega and the planks from its lumber mill in Arkhangel'sk, which helped them to survive the year when fishing was suspended and the farm suffered losses.

All of the timber harvesting of the kolkhozy is carried out in the so-called rural forests, which previously belonged to the farms but in 2004 were transferred to the control of the regional authorities and began to be divided up at auctions and based on tenders. Since fishing kolkhozy were unable to compete with major timber companies or even small specialized timber enterprises for a raw timber base, in the last few years it has become increasingly difficult for them to get wood for their activities and the new Forestry Code passed in 2006 will make it even more difficult to lease forests. Thus profitable timber harvesting is also a relatively new strategy and fits entirely within the path previously taken. Here, the state, in supporting the major timber companies in this field with legislation, is squeezing the *kolkhozy* out of this area of activity.

Other types of extra new activities in kolkhozy, such as harvesting seaweed, involve low prices for the raw materials and difficulties with delivery and distribution. The development of other forms of activity in the majority of kolkhozy which have free-standing diesel electricity stations is also problematic due to the high cost of electricity and, as a result, the high cost of any production. Because of this, one kolkhoz set up a fish farm near Arkhangel'sk, but the kolkhoz is situated on the Onega Peninsula, far from this location. Thus the fish farm is another economically profitable enterprise owned by the kolkhoz, but is separated from it geographically. The workers at the fish farm come from Arkhangel'sk, meaning that the farm cannot provide jobs for kolkhoz members, but it can provide money to support life in costal communities.

An essential function of the kolkhoz is the support of unprofitable and completely inefficient production in Pomor villages. This reflects the kolkhoz's social, rather than economic, orientation and demonstrates that it is a social as well as an economic means of organizing the local community. Unprofitable production in coastal villages, which includes production of electrical energy, agricultural production, and coastal fishing, is directed toward maintaining the community, creating jobs, and ensuring self-sufficiency. We will focus on coastal fishing, which is self-sufficient, yet does not make a profit for the collective farm. Besides providing jobs and fish for the community, support for coastal fishing is also support for a traditional way of fishing that is connected with the Pomor identity. Since, as we have already noted, deep-sea fishing is no longer part of traditional natural resource use and is even alienated from it, traditional fishing practices are preserved only in coastal fishing for herring, navaga, salmon, hunchback salmon, and flounder. One of the respondents said:

> Any fish that lived in the sea, we caught. In my day, we caught navaga and herring, but earlier they caught korekha, herring, flounder, whitefish, and bull trout. We fish from Kiy Island to the Konyukhovoy Gulf. That is, along the entire shore (Head of a fishing kolkhoz, m., b. 1963).

In addition, the traditional gear, methods and collective form of fishing are still being used. The same respondent said:

> But for the kolkhoz, we fish only with nets. We have about 80 nets for six people. It takes all day to collect them, and then we have to sort them out. After all, not only navaga are caught in these nets. Fluke and korekha also get in there. We

> dump them out onto the shore, which we freeze for the purpose. There we stir them around with rakes, so that they freeze, but do not freeze through. Next, we get down on our knees and sort them, small fry with the small fry, and big ones with the big ones. The navaga have to be hand picked, large, straight, and frozen; this is how we produce our product.

Horses are used to carry the fish from the ice, and the brigade of fishers lives in fishers' huts during the entire catch. These round-timber huts are about 4 metres by 3 metres, are lit with kerosene lamps, and have a small stove with a cooker. Six people sit and sleep on plank beds that surround a table.

The sale of fish from the coastal catch is arranged by the buyers themselves. A different respondent explained:

> Certain people concern themselves with the fish. They call, they find out if there are fish or not; they come, buy them up and sell them in Arkhangel'sk, Severodvinsk, and Novodvinsk. Today, they come from close by. In about two and a half hours they drive here, carry away the fish, and sell them back there. They drop by, find out how many fish have been caught, drive to the kolkhoz, pay their money in the office, get their invoice, and return here. It's all the same to them, since on the return trip they drive past anyway. When we didn't have roads, we took the fish on horses to the southwest logging road. There, we had to arrange the time that the car carrying the buyers would arrive (Head of a fishing kolkhoz, m., b. 1955).

Payments for biological resources, in this case fish, are high enough, but as long as the kolkhoz withholds certain deductions, it pays out only part of the money, and the fishing remains self-supporting.

Traditional coastal fishing is also preserved when Pomors fish for themselves. Everyone takes part in these catches – the men, the women, the small children, and even the very old. They look upon the catch as an integral part of their livelihood and as a traditional activity without which they could not imagine themselves living and through which their coastal identity is constructed. One of the respondents said:

> When the ice appears, the old women come out onto the ice to catch navaga. We sometimes quarrel over our spots. We all want to eat the fish, and only those that can't don't go out (Pensioner, former finance manager of kolkhoz, f., b. 1931).

The traditional, specially ornamented fisherwoman's jumper, worn only for fishing, proves that among the Pomors fishing was traditionally not only a male occupation, but also a female one.

Today, life on the fishing kolkhozy is hard enough. However, personal life strategies are frequently connected with the collective farm. This is true for the young and for the middle-aged.

> Yes, most often now the young remain, because there is nothing for them to do
> in the city (Head of a fishing kolkhoz, m., b. 1967).

The young people used to move to the city and find work. Today, work has become significantly more difficult to find. Even more so, it has become more difficult to get an education. If at one time there was a system of free elementary and higher education, now it has practically turned into a system that requires payment. However, kolkhoz workers earn only two to three thousand roubles a month – lower than the regional subsistence wage – and cannot pay for their children's higher education.

In the Soviet Union, village teachers were mainly outsiders. The system distributed graduates of teachers' colleges among the villages. New graduates, being young specialists, had to work three years in such villages. No such system exists anymore, and teachers, including those who used to live in the villages, try to find work in the cities or suburbs. In coastal villages, teachers are often graduates from village schools who do not have any special education. As a result, the level of teaching is low, thus making it impossible for graduates to move onto college. The same respondent said directly:

> If the children are not educated, they can't get situated in the city, and they stay
> in the country (Head of a fishing kolkhoz, m., b. 1967).

Other everyday problems could also be added to this picture. Among these are furnace heating; the absence of sanitation; low levels of electrical power resulting from outages and the eight to ten hours per day that the diesel electric plant works; one television channel – that does not always work; very bad telephone connections; the absence of mobile phone service and internet connections; and lack of access to qualified medical care, legal services, cosmetic services, and other services characteristic of the city environment.

Low wages, the everyday village routine and the loss of the traditional holiday culture during Soviet period all prompted parents and children to conclude that 'there is nothing to do in the village'. From this picture, it is clear why village youth are trying to leave their native regions and migrate to the cities.

Collective fish farms understand this mentality and are trying to help the young make their choice. A respondent said:

> Fortunately, in general, they find a place on the kolkhoz, this one at the diesel
> station, that one at the local course for tractor drivers. We train them. We disburse
> as much money as possible for our children's education in colleges and technical
> schools. If someone wants to do work that doesn't fit our profile, we give him a
> loan, and he slowly pays it back after he starts working. If the work does fit our
> profile, it is connected with the fishing industry, then he receives the education
> for free. This comes from the fishing profits (Head of a fishing kolkhoz, m., b.
> 1967).

We turn to a vital strategy that already extends beyond the framework of the fishing kolkhoz and by which the local community is trying to organize its life. This strategy, which is closely connected to the coastal identity, consists of creating and registering a native minority community. It can be considered an attempt to move off the beaten path (path dependency); that is, it is the creation of a new social institution in the coastal community.

Where the formation of the Pomor identity is concerned, on the one hand, native residents of the fishing villages are called and call themselves Pomors; on the other hand, since such a nationality did not exist in the Soviet Union, they were registered as Russians. This situation began to change in the 1990s, when there occurred a surge in national self-consciousness.

At that time, several Pomor organizations were established in Arkhangel'sk. These organizations not only revived the culture, traditional holidays, clothes, songs, and so on, but also achieved recognition of the Pomors as a native minority and their inclusion in the list of nationalities.

This was the most that they could do. However, these organizations were not able to build a connection with the Pomor villages, because the Pomor identity is formed in the villages with little or no input from the Pomor organizations. In addition, the village population knows practically nothing about these organizations. One of the respondents said:

> This question is far fetched. When we had a population census, I personally harboured great resentment toward our administration. They didn't even tell us that we could register and declare ourselves Pomors. And, in that all-Russian census we were simply designated as Russians. Not only in one such village, but everywhere. If we had known that we could have, we would have registered as Pomors everywhere. I know that I am a native Pomor. Maybe they would treat us differently, as they did with other nationalities. In my opinion, the Arkhangel'sk Pomor organizations exist for themselves alone. They've spent no time in Pomor villages, and they're not interested in us (Village teacher, f., b. 1958).

However, residents believe that Pomor organizations will work in the country as well as in the city, and the true census will finally take place. In this census, Pomors will be able to freely declare themselves Pomors. Another respondent said:

> The true census will come. After all, there are also Pomors in Arkhangel'sk, in Severodvinsk; how could there be a Kholmory without Pomors; in Mezen, they're all Pomors (Hunter and fisher, former teacher, m., b. 1960).

In addition, residents of coastal villages understand that their Pomor identity is formed by their connection to the sea and that, above all, residents of coastal villages can call themselves Pomors. The same respondent said:

But after all are people in Arkhangel'sk really as closely connected to the sea as we are? We are connected to this sea by a living umbilical cord.

Many village residents and the management of the fishing kolkhozy understand that their recognition as an indigenous minority would give them a chance to preserve traditional natural resource use. The local population, which has always fished with traditional gear for themselves and for the collective farm, does not understand why several years ago they introduced a charge for using biological resources, that is, for their traditional catch. The introduction of this fee put the local population on the same footing as any city-dweller who comes to catch fish on the seacoast. In conjunction with the changes in the law, a whole 'war' broke out when armed men came with the fish inspector and took the 'illegal' fishing gear from the Pomor fishers. After a while, this situation was rectified, but as we noted above, it could change at any time, and the collective farm members will have to pay the full fee for use of biological resources. Residents think that having the status of indigenous minorities would enable them to avoid such a situation. A respondent said hopefully:

Maybe if they had registered us as Pomors back then, they wouldn't have stolen our nets, and cut them loose. Here everyone, all the fishers, had their nets cut loose. They decided down there that we drain the sea's biological resources. It was such an insult to one's nets cut loose, and later in the same yea, they spilled oil into the sea. But the sea is our life (Hunter and fisher, former teacher, m., b. 1960).

Understanding this situation, some of the kolkhoz chairs began to actively support links with Pomor organizations, and one of them initiated a Pomor community organization in the villages belonging to his kolkhoz. The initiative was supported by the collective farm workers, and the community was registered as a social organization. A respondent reported:

We've formed an indigenous minority community with three villages and an island. However, we literally are just beginning our activity. All this turns on the fact that people want to live today and not sometime in the future. It turns on the fact that we will be able to get biological resources and have priority to take them, only if the government recognizes us (Head of a fishing kolkhoz, m., b. 1967).

Even in this case, however, the Pomor community is afraid that problems will emerge. The bureaucrats could allow them to catch a certain quantity of fish for food, but not to sell. The same respondent said:

And after all there is the 1982 Convention on Maritime Law, which talks about the economic independence of coastal communities. We don't need a primitive life, eating only fish, but we need to survive economically.

As a result, this Pomor organization is already more representative than the *kolkhoz*, and, if needed, can help the population stand up for their rights more effectively. For example, this community is close to the leased base of OAO Onegales, which is certified according to the international system for forest certification FSC (Kulyasova and Kulyasov 2007: 23–27). The certification standards stipulate that indigenous minorities have a priority right to exploit the forest, that they should be consulted, and that they have control over the plan for wood harvesting. Moreover, an organization such as the community can sign contracts for cooperation with Onegales (Karpachevskyi and Chuprov 2007). The negotiations between the community and Onegales have already started. In this way, the strategy for the creation and registration of the Pomors communities already might be successful with the defence of the nationality's rights to traditional forest use. When Russia acknowledges the Pomors as an indigenous minority an already prepared organization will defend the right to use the sea's resources.

Conclusion

The analysis of the transformation of nature management in the Pomor community showed that the Pomor village as a community presents itself altogether as a stable set of daily and seasonal practices. The social institutions of the Pomor villages maintain stability and adherence to the beaten track (path dependency). The fishing kolkhoz functions as a fundamental social institution. The first fishing kolkhozy included features of both pre- and post-revolutionary fishing cooperatives. The Pomor fishing kolkhozy preserved specific sociocultural features, connected with traditional culture and traditional nature management.

The appearance of the fishing kolkhoz, in the late 1920s and in the 1930s, became the most significant transformation of the community in the fishing villages for the entire twentieth century. In this period, the Soviet power made great efforts to eradicate traditional society and traditional nature management and aimed to make nature management part of the exclusive utilitarian economic trail. In this way, the leave from the beaten track of Pomor villages' community and their move to another form of development came about in the 1920s and 1930s and was connected with the policies of the Soviet government. New Soviet institutions shaped the path of Pomor society and built a new social structure.

Modernization processes took place in the 1950s and 1960s with the aim of changing the fishing kolkhozy into commercial industrial businesses based on trawling. New practices were introduced, but little changed in the kolkhoz as a social institution. We may say that new practices appeared inside the basic social institution that did not lead to a change of the beaten path or to a departure from the habitual track of social practices and that secured reproductive institutions. Despite a complete change in fishing technology in the course of several decades, its members and the amount of fish caught, the kolkhoz and its social practices of nature management were preserved inside the Pomor villages.

The modernization of some nature management practices such as ocean fishing did not lead to a change of the basic forms of activity and social institutions of the Pomor villages. The functioning of the fishing kolkhoz stayed on the beaten track and built new forms of nature management. But the development of the Pomor fishing village and the kolkhoz as its basic socioeconomic institution was disrupted by the reforms of the 1990s and suffered profound changes in the 2000s. Using the example of a kolkhoz, we traced how the organizational forms of the community and of nature management that developed historically in the pre-Soviet and in the Soviet times have been preserved in the current period of reforms.

The stability of the kolkhoz as a social institution in Soviet and modern times can be explained by economic and sociocultural reasons. On the one hand, the basic activity of kolkhozy – fishing – accounted for self-sufficiency to a large extent. Indeed, throughout all the Soviet period, the fishing kolkhozy were self-supporting. They had only their own resources to invest in new technology and infrastructure and they were able to support themselves with basic vegetables, meat, dairy products and fish. Therefore the economic reforms could not divert the development of the Pomor village onto a path other than the beaten track. The basic social institution of the village was preserved thanks to the presence of organizational forms in it that were necessary to adapt to the reforms. On the other hand, their structure and activity corresponded to the historically developed forms of collective nature management.

The current stage of transformation of the Pomor community proves the possibility of a new crucial change. The government has radically altered the rules of the game in the essential areas of nature management such as fishing and forestry. It changes the laws in order to strengthen large business in the field of fishing and forestry. This directly affects the life of the Pomor community and undermines the foundations on which it is built. The government pursues policies aimed at the diverting the fishing kolkhozes from the beaten track. However, looking at the example of the strategies analyzed, we can see that the community chooses strategies promoting its preservation in the context of the beaten path (path dependency). These are connected with the attempt to reinstate the traditional Pomor identity and to create new institutions that protect the traditional way of life.

We single out the following for the modern life strategies of the Pomor community: the preservation of the kolkhozy and the creation of a community of indigenous minorities and a community council. The kolkhoz collective, supporting all the village communities, made a conscious choice and stayed on the beaten track. The citizens of the villages supported the kolkhozy because they want to save their Pomor identity, their specific culture and their way of life. They decided to go the habitual path (path dependence). For the kolkhoz we also single out several new strategies that nevertheless had their beginning in the pre-revolutionary and early Soviet period: foreign partners, commercial forestry management and other types of activity yielding profit.

These strategies secured the efficacy of the kolkhozy and the preservation of the traditional way of life in the villages during the reform period. However, they may be less effective for the future. The government treats the kolkhozy merely as economical institutions, not considering their social and self-supporting character. Therefore the kolkhoz will be further forced out from the economic field of activity by the government. But this government strategy is not socially responsible, because the state did not suggest any social alternative to the people from these communities. The state operates according to the economic logic of big business and tries not to seeing the community's problems. It does not hear the voice of the Pomors or respect the choice of traditional life style by the Pomor community. Instead of the state, kolkhozy try to play the role of social bodies for the community and can do it effectively if they have a large enough fish quota from the state. If the state gives the Pomors the status of an indigenous people, it could be one form of social support and a sign that the state regards the Pomor people as having special cultural identity. Instead of pursuing its current policy, the government should try to protect these people as a minority with a unique culture and because they have no social dividends living in the very difficult conditions of the North.

The new strategy of the Pomors is the attempt to create a new social institution in the Pomor community and shape the beaten track (path dependency). This new institution is the Pomor neighbour community of indigenous minorities, which have NGO status. This strategy gives the community a new chance for the preservation of the traditional community and unique traditional culture.

References

Arkhangel'skaya regional'naya Territorial'no-sosedskaya obshchina korennogo malochislennogo naroda Pomorov, Pomorskaya Storona. Kul'tura i kratkaya istoriya Pomorov, korennogo naseleniya russkogo severa (2004) (Arkhangel'sk Region Territorial-Neighbourhood Society of Indigenous Minorities of Coast-Dwellers, Pomorian Side. Culture and Short History of the Coast-Dwellers, the Indigenous People of Northern Russia) (Arkhangel'sk: Izdatelskiy Dom).

Averkiev, A.S. and Chilin, M.B. (eds) (2004), *Vykhod iz Labirinta. Pazvitie metodov kompleksnogo upravleniya pribrezhnykh territoryi Kandalakshskogo raiona Murmanskoi oblasti* (Exit from the Labyrinth. The Development of Methods of Complex Management of Costal Areas in the Kandalakshki District of the Murmansk Region) (St. Petersburg: Russian State University for Hydro-Meteorology, UNESCO).

Auzan, A.A., Radaev, V.V., Nureev, R.M., Nayshul, V.A. et al. (2006), *Collection of Summarized Articles from the Symposium 'Path of Development: Continuity and Intermittence of Social Development'* (http://ecsocman.edu.ru/kir).

Bershtam, T.A. (1978), *Pomory. Formirovanie gruppy i sistema khozyaistvovaniya* (Formation of the group and the system of economic activity) (Leningrad: Institute of ethnography and anthropology).

Bulatov, V. (1999a), *Russkiy sever* (The Russian North) Vol. 1. *Zavoloch'e.* (Arkhangel'sk: Pomor State University).

Bulatov, V. (1999b), *Russkiy sever* (the Russian North) Vol. 3. *Pomor'e*. (Arkhangel'sk: Pomor State University).

Denisova, I.M. (2004), *Russkiy Sever. Etnicheskaya istoriya i traditsionnaya kul'tura XII-XX veka (*Russian North. Ethnical history and traditional culture XII-XX centuries) (Moskva: Nauka).

Filatov, V.P. (1994), Zhivoi kosmos: chelovek mezhdu silami zemli i neba (Living cosmos: A man between the forces of the land and sky), *Voprosy Filosofii* No 2, 7–12.

Hausner, J., Jessop, B. and Nielsen, K. (1995), 'Institutional Change in Post-Socialism', pp. 3–44 in Hausner et al. (eds).

Hausner, J., Jessop, B. and Nielsen, K. (eds) (1995), *Strategic Choice and Part-Dependency in Post Socialism. Institutional Dynamics in the Transformation Process* (London: Edward Elgar Publishing).

Karpachevskiy, M.L. and Chuprov, V.A. (eds) (2007), Rossiiskyi Natsional'nyi standart dobrovol'noi lesnoi tsertifikatsii po sheme lesnogo popechitel'skogo soveta. (Russian National standard of voluntary forest certification according Forest Stewardship Council scheme) (Moscow: Russian National Initiative of Forest Stewardship Council) http://www.fsc.ru/pdf/RNS.pdf.

Kochegura, A.T. (2001) Ot *Kolkhoza to Holdingy* (From Kolkhoz to Holding) (Moskva: Hauka).

Korotaev, V.L. (1998), *Russkiy Sever v kontse 19-go pervoi treti 20-go veka. Problemy Modernizatsii i Sotsial'noy Ekologiy* (The Russian North from the End of the 19th to the First Third of the 20th Century. Problems of Modernization and Social Ecology) (Arkhangel'sk: Pomor State University).

Krysanov, A.A. (2002), *Promyshlennyi lov treski onezhanami na Murmane (1850–1920)* (Industrial Catch of cod of the people of Onega in Murman) (Onega: Izdatelstvo Onega).

Kulyasova, A. and Kulyasov, I. (2007), 'Konstruirovanie vzaimodeistviya 'bizness – grazhdanskoe obshchestvo' na primere Onezhskii LDK / PLO Onegales' ('Construction of interaction between business and civil society, Oneshkiy LDK/ PLO Onegales case study), *Lesnoi Byulletin* 2 (35) June, 23–27.

Maksimov, S.V. (1984), *God na severe* (One year in the North) (Arkhangel'sk: Pravda Severa).

Mangataeva, D.D. (2000), *Evolutsiya traditsionnykh system zhyzneobespecheniya korennykh narodov Baikal'skogo regiona* (Evolution of the Traditional Systems of Life Support of Indigenous People of the Baikal Region) (Novosibirsk: Sibirskoe Otdelenie Rossiyskoi Academii Nauk).

Moseev, I.I. (2005), *Pomory Govoryat. Kratkyi Slovar' Pomorskogo Yazyka.* (Pomor Sayings. Short Dictionnary of the Pomor Language) (Arkhangel'sk: Pravda Severa).

Novikova, N.I. (ed.) (2003), *Olen vsegda prav. Issledovaniya po yuridicheskoy Antropologii* (Olen is always right. Research on Legal Antropology) (Novosibirsk: Izdatelskyi Dom Strategiya).

Popov, G.P. and Davydov, R.A. (1999), *Murman: Istoricheskie Issledovaniya XIX – nachala XX veka*. (Murman: Historical Studies of XIX – beginning of XX century) (Ekaterinburg: RAN, Ural Branch of North Ecological Problems Institute).

Terebikhin, N.M. (1993), *Sakral'naya Geografiya Russkogo severa* (Sacral Geography of the Russian North) (Arkhangel'sk: Pomor State University).

Titova, G.D. (2006), *Economicheskie Problemy ispol'zovaniya Biologicheskikh Resursov Mirovogo Okeana na territoriyakh Natsional'noi yurisdiktsii* (Economic Problems of the Use of Biological Resources of the World Oceans in the Areas of National Jurisdiction) (St. Petersburg.: VVM).

Vedenin, Y.D. and Kuleshova, M.E. (2001), 'Kul'turnyi Landshaft kak ob''ekt Kul'turnogo i Prirodnogo Naslediya. (The Cultural Landscape of as an Object of the Cultural Legacy)', *Izvestiya Rossiskoi Akademii Nauk, Seriya Geografiya* No. 1, 7–14.

Vlasova, I.V. (1995), 'Naselenie tsentral'nykh raionov russkogo severa XII-XX (The Inhabitants of the Central regions of the Russian North XII-XX centuries), *Etnograficheskoe obozrenie* 2.

PART III
The International and Global Impact on National Environmental Policy and Local Forestry and Fishery

Chapter 11

Local Adaptation to Climate Change in Fishing Villages and Forest Settlements in Northwest Russia

E. Carina H. Keskitalo and Antonina A. Kulyasova

Introduction and Theoretical Framework

Climate change will have a large impact on community level. As the emissions causing climate change are unlikely to be halted entirely in the near future, local areas and resource users may expect to have to adapt their ongoing land use to some extent (Ford and Smit 2004). Climate change is projected to lead to a more temperate climate in northernmost Europe, resulting in a delayed autumn and milder winters with increased precipitation. This change in climate, and adaptation to it, will cause an additional stress on resource users. The winter season will be shortened by perhaps a month (Hogda et al. 2001; ACIA 2004) and the ice on rivers and lakes will break up earlier and freeze later (IPCC 2001). This in turn will impact the spring flood and the seasonality of the flood may change (Hogda et al. 2001; IPCC 2001). In summer, plants may suffer from heat stress and soil moisture deficit due to the longer growing season and changing precipitation patterns and be exposed to invading pests, pathogens and herbivores. On the other hand, growth conditions may improve as a result of warmer temperatures and the ground being snow-free longer (Krankina et al. 1997; Saelthun 1995; Watson et al. 1998; cf. Maracchi et al. 2005; ACIA 2004).

Such changes could be expected to have the earliest effect on those parts of communities who are most directly dependent on the environment. The concepts of vulnerability and adaptive capacity have been used in research on both hazards and climate change to describe a community's susceptibility to change and possibilities to adapt to it. Vulnerability has been broadly defined as the 'capacity to be wounded' (Kates et al. 1985, p. 17). It is a measure of the exposure sensitivity (that is, how much a stress like climate change is expected to affect an area, given the exposure to the stress and the sensitivity of systems to it), less the adaptive capacity of socioeconomic, political and environmental systems to cope with or respond to this stress (Smit et al. 2000). Adaptive capacity can be defined as the potential for coping and adjustment in a societal system that allows for particular adaptations (Smit and Wandel 2006). In recent literature, vulnerability has been seen as largely targeting social vulnerability, that is, a focus on that it is the existing resource situation and to a large extent the adaptive capacity that determines the extent to which a community or other socio-economic or political unit is impacted by or can respond to change (Adger 2006). Social vulnerability thus argues that the access to resources such as political, economic, and social entitlements and resource rights play a large role in determining vulnerability and adaptive capacity at large, in response to whatever exposure to hazard may occur (Adger 2006). The concept of adaptive capacity also highlights that different communities even in the same area may be impacted differently by a stress, depending on the extent to which they can cope with or adapt to it (Smit et al. 2000; Smit and Skinner 2002). The degree of adaptive capacity can be seen as defined by a number of general determinants that may interact to different extents depending on the particular characteristics of the case and localities being studied (cf. Eakin and Lemos 2006; Smit and Pilifosova 2001; Moench and Dixit 2004; Cutter et al. 2003). These include economic wealth, technology, information and skills, as well as infrastructure, which embraces the availability of and access to resources by decision makers, institutions for, among other things, the distribution of resource rights, and the equity of resource distribution (Smit and Pilifosova 2001).

In this chapter, vulnerability will be applied as an overarching concept denoting the existing resource use and access situation in a community given existing stresses, thus focusing on social vulnerability, while climate change will be seen as a specific exposure-sensitivity or stress that affects and may add to that vulnerability (cf. O'Brien and Leichenko 2000). Adaptive capacity is seen as the capacity for coping with and adapting to stresses within an area, community or sector determined by broad community and societal characteristics. What is relevant here is to assess the degree to which the general socioeconomic and political capacities to deal with stresses are available to communities at the local level, as well as to define the level of governance – understood as the decision-making network of both public and private actors – on which adaptive capacities are determined (cf. Keskitalo 2008). A focal concern is whether actors are able to adapt locally within a community, or whether adaptive capacity mainly resides with the state or other frameworks. Such a distinction is important as the changes that require access to, for instance, legislative or other system changes would be dependent on access to external actors rather than undertaken directly on community level, with potential impacts on adaptive capacity and vulnerability. Specifically, the study addresses the following questions:

- What is the current socioeconomic situation and in what ways, and to what extent, will climate change as described in the impacts and scenario literature have an effect on communities?
- What are the determinants that are most important for adaptive capacity in the case study, and at what levels are adaptive capacity situated?

The study aims to describe the effects of change and the capacity to adapt to combined stresses as seen through the eyes of local people, i.e. as narratives of perceived vulnerability in a current situation and the capacity to adapt to both existing stresses and climate change. Wide-ranging change can be expected to have the most detrimental effects on those who have the least resources and are least able to draw upon economic and political networks for support or adaptation within a market or legislative framework (O'Brien and Leichenko 2000). The study targets the case of fishing villages in the Arkhangel'sk coastal area and forest settlements in the Arkhangel'sk Region. The choice of case study area and sectors is based on the fact that renewable resource industries and users are the most directly impacted by climate change, as climate change may increase natural variability and make resource availability less predictable. The local communities studied are to a large extent dependent both economically and – given the structure of social organization in Russia – socially on fishing cooperatives or logging enterprises. Local communities also depend on the use of other renewable natural resources – gathering berries and mushrooms, hunting, fishing, and using firewood for heat and wood for building. These factors as well as the peripheral location of many of the settlements in relation to regional and national decision-making could be expected to cause a heightened vulnerability to changes in both governance and resource rights distribution.

Methodology and Case Study Areas

The forest industry traditionally plays a large role in the Arkhangel'sk Region (*oblast'*) in Northwest Russia. Most of the rural settlements in the region are forest settlements. The fishing industry and fishing cooperatives (*kolkhozes*) also have considerable importance in the region, especially socially and culturally, as they support coastal villages with the traditional Pomor culture, which is connected with fishing. This study mainly utilizes semi-structured interviews with stakeholders in local communities that are part of the fishing kolkhozes and logging enterprises (*lespromkhozy*). The former are located on the White Sea coast near Arkhangel'sk and in the Onega and Primorsk districts on the Onega Isthmus, and include the kolkhozes Red Banner (Krasnoe Znamya), Forty years of October (40 Let Oktyabrya), In Lenin's Name (Imeni Lenina), and Magus on the White Sea Coast (Belomor); the latter operate in the Arkhangel'sk Region and the Onega, Kargopol and Plesetsk districts (the logging enterprises Onegales, Kargopol'les, Yarnemales) (see the map at the beginning of the book). The study thus centres on conditions for fishing, logging and subsistence in rural areas. The qualitative and open-ended design centred on semi-structured interviews is intended to ascertain the understanding of large-scale change through the interviewees' rather than outsiders' eyes (Keskitalo 2008). The aim is to provide an assessment of the perceived changes and stresses that impact social vulnerability, as well as an understanding of the extent to which climate or environmental change impacts renewable resource and subsistence users and what determines their adaptive capacities. The interviews were mainly undertaken during the period 2005–2006 and include in total about 55 persons between the ages of 30 to 86 working in fish kolkhozes, logging enterprises and local branches of state forest agencies (leskhozy), pensioners, representatives from local administration, hunters and fishers, as well as people who were registered as unemployed and who make their living from individual subsistence farms. Within the collective farm system described above, interviewees were mainly selected from among people who had worked for a long time – as much as a generation, or most of their lives – in the areas. This means that the stakeholders interviewed have lived through and can describe changes in the economic, organizational/ political and natural environment over time.

The issue of climate change was mainly discussed by presenting the interviewees with statements on broad climate change impacts phrased in lay terms, the aim being to focus on stakeholders' understandings of change rather than on the uncertainties inherent in determining levels of climate change. Thus, questions were phrased not to include the term 'climate change' but to ask about changes in the weather and nature instead. The interviews were structured as follows. Firstly, interviewees were asked to describe the structure of their work or livelihood in broad terms, how changes they have seen in their livelihood have impacted them and what they considered to be their main problems and prospects. They were also asked to describe their possibilities to adapt to these changes and, where these existed, the means available to them for doing so. These questions were posed in order to view general social vulnerability to changes at large, aside from climate as a focal stress. Secondly, stakeholders were asked if they had perceived changes in the environment, and if so, which, as well as were asked a number of questions regarding potential or actual impacts of changes that could become recurrent with climate change. These questions were expressed in a lay format (for instance, 'How would it affect you if spring came earlier?'), and derived from a survey of climate scenario and impacts literature regarding general impacts predicted for the region (IPCC 2001; Hogda et al. 2001; Krankina et al. 1997; Watson et al. 1998; Maracchi et al. 2005; ACIA 2004; Saelthun 1995). Specific questions regarding climate change impacts and adaptation thus concerned what changes, if any, the interviewees had seen in the weather and nature during their life, and how it would affect them if there were warmer winters with more snow and thaws, autumn came later, or spring came earlier with thawing and re-freezing and the time of the spring floods changed. Interviewees were also asked how it would impact them if summers were warmer and drier, forest growth increased, there were more pests or forest fires, and the weather were less predictable, with more storms. Interviewees were also asked what they could do to adapt if these changes were to take place.

The sections below describe, firstly, the social vulnerability of areas with reference to their resource access and organization. The section after describes the added impacts of climate change given this situation, and the final section discusses the main determinants of adaptive capacity in the focal case.

Socioeconomic Situation and Resource Access: Social Vulnerability

The socioeconomic conditions and resources of the fishing villages and forest settlements are important in evaluating their potential for adaptation to social and climatic modifications. Social vulnerability can be seen as a result of the historical development and infrastructure in the fishing villages and forest settlements, as well as access to fishing, forest and subsistence resources.

Historical Development and Infrastructure in Fishing Villages and Forest Settlements

Historically, fishing kolkhozy were organized as part of Soviet state policy in the 1930s as a cooperative form of village economy and were reorganized in the 1950s to include several neighbouring villages. Kolkhozy were thus based on traditional, often Pomor, settlements and local decision making. Kolkhozy were also inherently connected with the social organization of the villages, with their infrastructure organized by the kolkhoz and funded by profits from fishing. This made it possible for remote villages to maintain collective-farm agriculture and infrastructure that, among other things, provided them with milk and meat. The kolkhoz was thus responsible for social services in the areas. The fishing kolkhoz was unique in Soviet times in working as a cooperative society which itself planned its own activities and use of profits rather than following state plans. Kolkhozy had a state plan specifying the amount of fish they had to give to the state annually, but they could meet this requirement as they saw fit. As a result of this independence, kolkhozy were generally not liquidated or incorporated into larger units following the fall of the Soviet Union; rather, they maintained their identity as local cooperative societies (see Chapter 10 of this volume).

In contrast to the fishing kolkhozy, Soviet state logging enterprises (lespromkhozy) appeared, for the most part between the 1930s and the 1950s, in locations that were convenient for logging, often where settlements did not previously exist; they followed state-determined logging plans and took responsibility for social services in the settlements that were established. The lespromkhozy were often made up of in-migrants from a number of places, including the southern republics of the Soviet Union. During the reforms following the fall of the Soviet Union in the 1990s, the lespromkhozy were converted into joint-stock companies, most of which finally became part of larger forest holding companies. Thus, unlike the fishing kolkhozy, the lespromkhozy, lost some of their connection to the local context in the 1990s; they may also have a more limited basis of traditional local knowledge that can support adaptation to socioeconomic and environmental changes (Kulyasova and Pchelkina 2004; Kulyasova and Kulyasov 2005; Kulyasova et al. 2005). Employment in forest settlements has also been severely reduced. The fishing kolkhozy haves generally attempted to maintain the social infrastructure of the villages, but at the cost of making it more difficult for the enterprise to rationalize its activities and compete on the economic market. In the case of forest settlements, however, enterprises have generally limited their investment in social life and infrastructure, although retained some role in relation to the corporate social responsibility of the holding company as a whole.

In both cases, the loss of social responsibility and infrastructure has had severe impacts on the local community and on access to electricity, telephone and mobile communication connections, as well as railway and automobile lines. In most of the fishing villages and in some forest settlements there today exists no connection to public electricity lines, but only mini power stations that run on oil. As a result of the high cost of oil and oil delivery, one kilowatt-hour produced by a local mini power station may cost as much as 10-20 times what it does when supplied on the general network in the countryside. In addition, the capacity of mini power stations is low and restricts electricity use to night time, when it is needed the most, effectively making day-time use of electrical household appliances and computers impossible. Thus, local logging enterprises cannot for example use electrical saw equipment, and fishing kolkhozes cannot use industrial refrigerators for fish storage. As a result, fish is stored outside in the winter, making storage reliant on below-freezing weather and subject to damage in warm winters or winters with frequent thaws. Telephone service exists between the majority of fishing villages and forest settlements but is slow and frequently hampered by weather conditions (high winds, rain, and breaking wires). In practice, there are no mobile communication connections; satellite communication connections are very expensive and essentially only available to managers of enterprises; access to the Internet is lacking.

Where transportation is concerned, most forest settlements have road or railway connections, but the roads are in poor condition and become worse especially in rainy periods. Fishing villages are in a particularly complicated situation, as most of them are only accessible by winter roads. During other times they use waterways or airplanes, which are very expensive and make it difficult to organize the transport of products economically. As a result, any shortening of the period during which there is stable below-freezing weather and winter roads are accessible will have an effect on the economic situation of kolkhozes and the related communities.

The absence of infrastructure, in both fishing villages and forest settlements, thus decreases socioeconomic adaptability and results in a major dependence on weather conditions. The conditions in the remote villages also result in a relatively large outflow of youth to the cities in search of work.

Access to Fishing Resources

The existence of enterprises depends on access to resources. Fishing kolkhozy undertake quota-based marine fishing from large trawlers and also utilize subsistence ice-fishing close to the shoreline (cf. chapter 10) during the period with thick ice; this has regularly been November to April, but in the last five years the season has been limited to the period from December to March or even January and February.

The kolkhozy pay a fixed fee to the state for this use of bioresources; the fee can be reduced if the kolkhoz can prove that it is the main enterprise supplying the community with social services. Fish harvested within quota-based fishing is most often sold for export, for instance, to Norway or other European states or to Asia, as export prices are higher than those on the domestic market. The fishing quotas are decided in the Norwegian-Russian Fishing Commission, with each state then distributing its national quota. Units such as the collective farms are awarded a certain quantity free of charge and to catch for their own consumption, but since 2001 have had to compete for additional quotas in auctions where the quota goes to the highest bidder. This in effect decreases the resources available to small-scale collective farms with limited economic resources, and the quota available to each collective farm may now be half of what it was prior to the auction system. Kolkhozy estimate that they require 1,000 tons per kolkhoz, while a trawler can readily catch 2,000 tons. There are also minimum limits on ship-specific quotas whereby a trawler must have at least a 600-ton quota to be allowed to fish.

Thus, kolkhozy have considerable problems in acquiring rights to marine fish resources. Because of limited economic resources, the collective farms are not able to get bank loans to support competition with companies for quotas, and are thus not able to gain access to resources that would enable them to subsist economically. As one interviewee expressed it:

> [It] turns out that if you have economically effective facilities, it means that your application will be considered and if this is not the case, then it will not. And the diesel station in our community is unprofitable, agriculture is unprofitable, and they [the banks] say that 'you won't be able to pay us back'. Therefore banks generally refuse us credit. They see that we have only 300 tons, they say that 'with these 300 tons we cannot give you a permit' … we are always turned down, and the reason is 'you are not solvent'. We say that we survive in difficult conditions, support the population, pay attention to our social importance [as well as] production efficiency. We try to show that we are not efficient not because we are … not able to work, but because we work under difficult conditions in the north.

As an economic adaptation, kolkhozes have united in associations in order to pool quotas and trawlers, as well as the financial resources for purchasing quotas at auctions. Many kolkhozy have received offers from large fishing enterprises to join joint-stock companies or holding companies, but have most often rejected these offers, as it would mean a loss of independence for the collective and a change in the role of the kolkhoz in the local community (cf. Chapters 5 and 10). As a result of these problems of resource access, some kolkhozes have also investigated the possibility of claiming indigenous rights based on Pomor traditions. Indigenous peoples are according to Russian legislation accorded larger resource rights than purely local groups, and if Pomor were to be recognized as an indigenous group, local users who could claim this background would be able to gain larger fish quota rights.

Access to Forest Resources

In the Soviet period, forest settlements were constructed in places where there were significant stocks of wood, exhausting much of the timber reserves in Northwest Russia; in Arkhangel'sk, however, there are still regions with old-growth forest, protected by international and Russian environmental organizations (cf. Chapter 9). While there exist major forest-covered territories in these areas, many of them are not accessible on account of the limited road network, and in certain areas long transport distances to consumers make logging unprofitable. Most logging also utilizes winter roads, which are roads that are only possible to use during the period when the ground is frozen. This situation means that any shortening of winter would impact logging negatively given the current infrastructure.

The competition for wood resources has also increased in the areas since the mid-1990s, when it became possible to lease forest plots from the state for logging following regional competitions or auctions. While the state owns all land, forest companies are thus able to lease plots for time periods ranging from a few years to several decades, with the result that only those who are able to compete at auctions are able to lease land. The price of roundwood at the end of the 1990s and in the 2000s was also comparatively low and resulted in that many logging enterprises operated at zero profit and could not pay wages and taxes or invest in new machinery. As a result, during this period many enterprises became parts of larger holding companies which included more profitable sawmills or pulp and paper mills. The redistribution of profits within the holding structure allowed for investments in logging enterprises and the stable payment of wages and taxes; however, these changes led to a dominance of large holding structures on the local level.

In the Arkhangel'sk Region, the general scheme of management for the enterprises in the holding company PLO Onegales has helped the enterprises pay stable wages and taxes and become more efficient by acquiring new technology such as harvesters and forwarders. However, under the new Forest Code adopted in 2006 all forest plots for logging in the Russian Northwest will be distributed by auction under relatively tough competition, where large-scale forest holding companies may win out on account of larger resources. This may have effects on communities should leaseholders log areas surrounding communities. To some extent, however, may community use of forests be supported by the fact that all settlements in the case study are situated near FSC-certified territories (see more on the Forest Stewardship Council, FSC, in other chapters in this volume; Kulyasova and Kulyasov 2005; Kulyasova, Kulyasov and Pchelkina 2005). Certification – the opportunity to label wood as certified when harvested according to certain environmental and social criteria – is seen by holding companies as a market requirement and is part of their policy in the area, among other things resulting in the holding companies paying for certification of their constituent enterprises. Certification has economic benefits for the company in that it gains access to the stable European market; the certification process also prompts companies to internalize environmental values and requires a certain level of corporate social responsibility such as retaining forest areas for local use and provision of firewood.

Village Subsistence Resources

While financial resources in fishing villages are largely dependent on fishing quotas and collective farm arrangements, regular day-to-day life in the villages is dependent to a large extent on subsistence resources, including forest and coastal resources. In forest settlements, daily life is dependent on forest, river and lake resources, as well as on subsistence farming. Fishing villages are especially reliant on local subsistence resources given the limited transport possibilities and low incomes in the kolkhozy. Many of the villages have small-scale agriculture maintained by the kolkhoz: they grow potatoes and vegetables and raise horses and cattle for milk and meat, primarily for subsistence use but sometimes for sale as well. The villages also harvest hay for animal fodder. The number of both privately owned and kolkhoz-owned horses has increased since the 1990s; the animals are used in agricultural work and for transportation, as petrol and spare parts for tractors are expensive and delivery times may be long and uncertain. Some villages also hunt seals and cultivate seaweed, although any prospects of increasing seaweed production are limited by the areas not having access to export networks for the resource.

In both fishing villages and forest settlements people also gather berries and mushrooms. Many people dry or preserve berries (cowberry, bilberry, cloudberry, cranberry) and mushrooms, which are used for subsistence or sale. The gathering of mushrooms and berries is common in Russia; it forms the basis of the budget of many families – especially those who are unemployed – and may be essential even for those who have other incomes. In all communities in this case study, the gathering of berries and mushrooms for sale at a storage point is a large-scale activity. The storage of berries is organized in villages and settlements, where local inhabitants are employed by corporations who purchase berries and transport them to larger storage facilities or cities. For families who engaged in berry picking during the entire season, the earnings from berries were comparable with the income from work at the logging enterprise; in many fishing villages, however, large-scale berry picking and sale was difficult due to lack of transportation. Hunting is also important, both as a source of meat and as a social practice. Hare, elk and forest birds are hunted to provide both fresh meat and salted meat for winter. Declining amounts of game and game diversity due to non-local hunters and ecological changes are, however, reducing the importance of hunting. Some villagers also own small boats (typically around six metres in length) for small-scale coastal fishing; however, most commonly villagers catch fish for food by ice-fishing. In total, any changes in the environment, weather or climate that affect the large-scale harvesting of berries or wildlife distribution will thus substantially affect the economic security of the inhabitants of forest settlements and fishing villages.

Local forests are also important for villages as a result of the use of local wood for heating and as building material. In fishing villages this need is partly met by the kolkhozes, which log and saw a small amount of wood, with the work being done partly by the residents themselves. In forest settlements, as a rule, the logging enterprises sell logs, often substandard logs that cannot be used for saw or pulp and paper mills, to communities as firewood. However, as logging technology improves there is less and less such surplus wood available for stove heating purposes; the distance from the cutting plot to the settlements has also increased with increased logging, which raises the price of firewood. In some cases the logging enterprises give permission to local people to log for their own needs on the leased territory. Both fishing kolkhozy and logging enterprises try to provide local social institutions (schools, libraries, houses of culture, medical offices and kindergartens) with wood for free or at low prices. However, the timely supply of wood for heating, building and repairing remains a problem. As stoves are used for not only heating but cooking as well, warmer temperatures (due, for instance, to climate change) will not necessarily significantly reduce the amount of wood needed in homes. Continued access to wood resources is also becoming more restricted, as the system by which the state sells resource rights to the highest bidder may encourage logging companies taking out as high amounts of wood as possible from forests, in order to gain returns to their investment.

Impacts of Climate Change

Given the extensive reliance on the environment for subsistence in the region studied, climate change may exacerbate the current situation in that it impacts resource access by shifting the times when resources are accessible, as well as the quality and quantity of resources and services such as winter storage and travel. Interviewees generally viewed projected climate change impacts as mirroring the changes seen in recent years, which suggests that climate change may already be ongoing. Villagers were thus able to point out the effects of existing or experienced situations, and to a large extent described existing responses. Overall, the changes they noted indicate that weather is becoming more unpredictable, affecting for instance the timing of different activities (such as transport or fishing dependent on ice conditions, or sowing crops). The longer period of open water will lead to a longer navigation period, meaning that water routes for transport will function longer in the winter; however, for villagers, having access to winter roads for travel by car was preferable to having ice-free waterways for transport by ship or boat.

Temperature Change and Seasonality

The interviewees generally stated that it has become warmer over time. Winter has started later, been milder and included thawing periods, and spring has also come earlier. This has made the weather less predictable. For instance, one interviewee noted:

> The weather for the last 10 years has become unstable. Earlier, if it was summer, it was summer. If it was winter, it was winter. And now [this winter] the snow already has melted three times.

A beneficial impact of the temperature change in winter is that somewhat less wood is used for heating purposes only if the weather is warmer. On the other hand, there are negative impacts on the accessibility of places that are only reachable by winter road, or by ferry or plane in the summer. For areas that are dependent on access by roads across frozen sea and river ice, access by frozen marshes is similarly limited, as the rivers, marshes and the sea all freeze later and thaw earlier.

A shorter period when waterways and the ground are frozen also has a strong impact on the local populations of fishing villages, for whom ice-fishing provides an important source of protein in winter. The varying and warmer weather both shortens the period for secure ice fishing and makes it more difficult to store fish and other foodstuffs over winter, as the villages and forest settlements that cannot rely on access to electricity use the outdoors as their freezer. Fish or other perishables are ruined if the thaws last long enough:

> Winter [comes] later and later every year. Earlier we caught fish for November seventh, for the holidays, and the sea was frozen. And now, look, every year it [is] later and later. ... [If] the ice is not present, also fish [are] not present; we do not catch them ... without ice ... And winters [are] warmer. In the winter when you catch fish, suddenly [there is] a thaw.

In order to adapt to thawing in winter, the collective farm Forty Years of October recently purchased its first freezer for the villages; a freezer is beneficial in that, unlike in a refrigerator, food does not have the time to thaw during breaks in the electricity supply.

Changes in temperature have also been seen in summer. Summers have sometimes become hotter with less rain. Many interviewees also observed that summers have become drier, with the risk of water levels in rivers falling, causing trees to dry up and limiting fish spawning and the amount of river fish.

Impacts on Fishing

For fishing activities, warming of the seawater has resulted in a somewhat different distribution of fish. One stakeholder with access to scientific data on temperature measurements and fish distribution noted:

> In the last three years fishers have noticed ... a change of average annual temperature of the water. The thing is that the fish very much react to it, even a change on the extent of a degree, on a tenth of a degree. And it is true, we have noticed, that in recent years the distribution, in particular, of cod, has extended, and already in those northern areas where it never occurred, now it can be seen.

Because of this non-conventional distribution of fish and the fact that the ice has receded to the north, the last three years have brought benefits to the fishing yield among trawlers. However, such changes in fish distribution may mean that coastal fishing, which uses smaller boats that are not allowed to transgress the 12-mile coastal limit for fishing, may not be able to access their target species.

> Previously, the migration routes were stable; now this is not the case. The fish move go to colder waters, to the north – the cod, haddock, the basic species of fish that we catch. For us, for trawlers, this is not of great importance, but [for] the coastal fishery [it is].

For instance, the head of one fishing collective noted that as quotas are defined for the different fish species (such as cod, haddock, halibut, or herring) and the vessels only carry equipment to catch certain of these, any change in the distribution of species could have a large impact. Interviewees also observe that storms and strong wind may become more frequent, further impacting small boats in particular. The navigation period may, however, be extended, as winter, which brings snow and freezing temperatures, will start later.

Additionally, the lack of ice was seen as impacting the possibilities for seal hunting, with seals potentially moving towards Norway, for instance, where they will not be accessible to local hunters. This is something that fishers worried may reduce fish resources further in the long run, as increased numbers of seals, not diminished through hunting, may consume more of the fish resources.

Impacts on Forestry

For forestry-related activities and for logging enterprises the shorter winter will have a large limiting impact on the access to logging areas through frozen winter forest roads. An increase in timber growth has been observed, one trend possibly related to warming. The major concern however, is that given the limited road network, logging enterprises and smaller collective farms will only be able to carry out logging during the period when winter roads are available, which in some years has lasted only from December to March, rather than, as earlier, from October to April. Some plots can be logged only when the ground is frozen, because marsh soil, moist soil, and weak soil are very susceptible to damage during logging under other conditions. Unpredictability around the length of the frozen period could also impact the possibilities for wood storage. If logs are harvested and placed by a winter road to be picked up for transport, they may be destroyed by insects and pests if there occurs a thaw that hinders transport. Machines such as tractors that attempt to transport wood may sink in marshland areas or bogs, potentially getting stuck and destroying the root system of trees. A warm winter would mean that many of the logging enterprises either could not operate on winter plots or would cause substantial ecological damage to the forest soil if operations were not ceased. A more unpredictable and shorter winter period would thus impact the possibilities for logging, transport and wood storage.

It was also noted that changes in temperature could have a large impact on the risk of forest fires. Interviewees noted that the risk of forest fires differ with conditions from year to year:

If June is damp, there are few fires. If it is dry, even if cold and dry, there are a lot of fires. Everything depends on the first month – May or June.

However, fires may have a lesser effect in undersized forests with relatively few cubic metres of wood per hectare than in other, denser forests.

Impacts from variation in temperature have also been observed on the main rivers in Northwest Russia, which are used during the spring flood for transporting logs. In the last few years, the spring flood has been extremely low because of the small amount of snow, and the logs that were prepared for water transportation were not floated down the river. On the other hand, interviewees noted that an unpredictably high spring flood could destroy logs stored on the bank. These varying observations are consistent with climate change impacts and projections literature that has noted variable levels of snowfall, snowmelt and spring floods (IPCC 2001). Bogs were also seen as having increased, possibly due to higher logging rates and greater snowmelt as well. At the same time, the number of dry areas at somewhat higher elevations has increased. An increase in the number of bogs was seen as further disrupting the possibility for communications between settlements. However, the formation of bogs may prevent problems with flooding, as marshes may absorb excess water.

Impacts on Subsistence and Invasive Species

Another result of warmer temperatures is that the timing for sowing crops in spring has been impacted. While earlier sowing would be beneficial for crop yields, there is a risk that recurring frosts might damage the crop and make it difficult to take advantage of a longer growing period:

> It is usual that snow melts in the middle of May and potatoes can be put into the ground. Now I put the potatoes in the ground already in May – the first or the second – but the thing is that there can be frosts at the end of May or beginning of–June; [there may be] temperatures such as [those] now [in winter] and in July it is again hot, warm; by September there is a downturn of temperature; in the first half of September, zero degrees, temperatures below freezing, frosts, and then sharp warming.

In 2006 there were unusual frosts during the whole summer:

> there were frosts in the beginning of June, in the middle and the end of June, and in the beginning and the middle of July; the potatoes, squashes, pumpkins were partly frozen.

There was also an unusual period of warmer temperatures in early spring and a long period of frost after and also a different pattern of rain in the warmer season (three to four weeks of rains and three to four weeks of sunny weather without precipitation instead of mixed weather). Such occurrences may make berry flowers freeze, impacting the crucial berry crops and also limiting mushroom harvests.

In addition, species not prevalent in the area earlier have been observed, such as vipers, which may be dangerous as people are unprepared for them. As one interviewee noted: 'snakes are thermophilic, they should [need to] keep warm … they never came here before'. Changes have also been seen in the behaviour of migratory birds such as swans and geese (which may however relate to losses of earlier migration sites and agricultural patterns of sowing), in the prevalence of ticks and in the occurrence of Colorado beetles and other crop pests not seen previously in the region. Cow-parsnip (*Sosnovskogo*), a very poisonous plant which causes skin rashes when touched, has appeared in several parts of the Arkhangel'sk and the Karelia Regions. The plant is an invasive species originating from the Caucasus and very aggressive, spreading very fast and replacing endemic species. Bees, previously rather rare in the Arkhangel'sk area, have also been seen more frequently, and one interviewee described how her cherry tree had suddenly started bearing fruit during a long, unusually warm summer that occurred in the last few years. In general, as one interviewee noted on the new species: 'These are kinds of animals and insects which lived in the south of our area, that is, somewhere 500–600 kilometres to the south; now they have moved here'. With regard to these changes, interviewees state that one problem in finding adaptive measures lies in the fact that there exist no traditional ways of responding to the problems:

> In the south people have already adapted for a long time …. but here nobody knows how to combat these setbacks [ticks, potato and crop pests].

Conclusion: Possibilities for Adaptation?

In general, many of the problems for the villagers can be defined as relating to the possibilities for maintaining an economy in the peripheral northern regions. The existence of a system where resource access beyond a certain level is distributed through a market (auction) system has a high impact on local enterprises and units, who are not able to compete as a result of their small size as well as of bearing much of the social costs of communities, as well as the costs of investing in new technology. This Catch 22-situation can be seen as leading to unemployment in many settlements. As a result, the governance or decision-making framework for access to resources (both natural resources and the terms for social services provision) that is set by the state and gives preference to larger-scale actors is a main determinant of social vulnerability and adaptive capacity (Smit and Pilifosova 2001; O'Brien and Leichenkko 2000). The main resource problems thus reside in establishing rights to supply of resources, rather than in demand for products given that sales networks and business connections exist. The general social vulnerability thus limits the possibilities for local adaptation overall and to climate change in particular.

An illustrative example of this pattern in the present case is fishing, which can be seen as highly vulnerable because of the economic requirement of supporting social services, which limits resources for adaptation. In fishing, large problems was seen to be the limited resources available to the communities, one being the lack of capital available to collective farms, which cannot obtain loans and thus lack sufficient resources to purchase quotas that would enable them to get access to electricity or improve the infrastructure. What could be defined as globalizing factors are thereby of considerable importance for adaptation (O'Brien and Leichenko 2000). To increase their profit margin and thereby adapt economically, collective farms sell their products for export at higher prices but are limited by internal resource conflicts and a regulative system that favours larger market actors. Legislation that restricts coastal fishing to within a 12-mile limit was also seen by interviewees as decreasing coastal fishing possibilities. The possibilities actors saw for potentially increasing their adaptive capacity over time mainly consisted of contesting the distribution of rights by claiming indigenous peoples' rights for Pomors. Any further combination of quotas or joining together into larger units may be difficult given existing problems and the market- rather than socially-oriented attitude in potential business partners. Many of the perceived problems and the adaptations suggested for dealing with these thus also here hinge on the regulative framework on the state level, which determines resource rights and thereby also economic opportunities (for instance, for bank loans in order to compete economically within the system). This impact of extra-local frameworks is consistent with patterns found in other research in northern areas (Keskitalo 2008), indicating that much of the adaptive capacity for localities are situated at and require actions at higher levels, thereby limiting the degree of adaptation that can be undertaken in the community.

The case of forestry and fishing communities in Northwest Russia further illustrates the higher vulnerability to climate change among local communities that are dependent on subsistence resources (cf. Ford and Smit 2004) compared to those where technology and well-developed physical and social infrastructures can be used to limit direct dependency on the environment. For the communities studied here, potential adaptation strategies such as technological change that may be available in more wealthy fishing communities as a response to climate change were not available. Examples of these may include, for instance, by cold storage, central heating, or focusing on technologies such as fish farming that would lessen dependency on fish stock fluctuations which have been seen in communities in northernmost Norway (Keskitalo 2008). Climate change may possibly exacerbate existing resource problems by limiting access to areas, thereby further increasing the reliance on electricity for cold storage and impacting local natural resource use (harvest and subsistence agriculture). In many cases, impacts such as those that could be expected with continuing climate change have already been perceived, such as changes in seasonality and the in-migration of invasive species. Adaptations to climatic changes may be less demanding when they relate to conditions in which variations have been seen for a long time and for which strategies exist, or in cases where they can be undertaken through individual-based adaptations, such as changes in crop patterns due to limited seasonal predictability. In some cases, adaptations may be able to include modifications to accommodate changes to unfamiliar conditions, such as those brought on by new species and for which adaptive or coping strategies do not yet exist. However, this adaptive range is limited by the institutional framework; more far-ranging adaptation on the local level would require changes on the level of state regulation involving resource access and infrastructure and other support.

References

ACIA [Arctic Climate Impact Assessment] (2004), *Impacts of a Warming Arctic – Arctic Climate Impact Assessment.* (Cambridge: Cambridge University Press).

Adger, W.N. (2006), 'Vulnerability', *Global Environmental Change* 16, 268–281.

Cutter, S. L., Boruff, B. J. and Shirley, W. L. (2003), 'Social Vulnerability to Environmental Hazards', *Social Science Quarterly* 84:2, June 2003, 242–261.

Eakin, H. and Lemos, M.C. (2006), 'Adaptation and the State: Latin America and the Challenge of Capacity-Building under Globalization', *Global Environmental Change* 16, 7–18.

Ford, J.D. and Smit, B. (2004), 'A Framework for Assessing the Vulnerability of Communities in the Canadian Arctic to Risks Associated with Climate Change', *Arctic* 57:4, Dec. 2004, 389–400.

Guisan, A., Holten, J. I., Spichiger, R. and Tessier, L. (eds) (1995), *Potential Ecological Impacts of Climate Change in the Alps and Fennoscandian Mountains. An Annex to the Intergovernmental Panel on Climate Change (IPCC) Second Assessment Report, Working Group II-C (Impacts of Climate Change on Mountain Regions)* (Geneve: Conservatorie et Jardin Botaniques de Geneve).

Hogda, K.A., Karlsen, S.R. and Solheim, I. (2001), 'Climatic Change Impact On Growing Season in Fennoscandia Studied by a Time Series of NOAA AVHRR NDVI Data', *Proceedings of IGARSS.* 9–13 July 2001, Sydney, Australia. Available at http://www.itek.norut.no/projects/phenology/en/results/publicat ions/IGARSS2001Hogda-et-al.pdf (Accessed September 9, 2003).

IPCC [J.J. McCarthy, O.F. Canziani, N.A. Leary, D.J. Dokken and K.S. White, eds] (2001), *Climate Change 2001: Mitigation. Contribution of Working Group III to the Third Assessment Report of the Intergovernmental Panel on Climate Change* (Cambridge: Cambridge University Press).

Kates, R.W., Ausubel, J.H. and Berberian, M. (eds) (1985), *Climate Impact Assessment: Studies of the Impact of Climate and Society* (Chichester, UK: John Wiley and Sons).

Keskitalo, E.C.H. (2008), *Climate Change and Globalization in the Arctic. An Integrated Approach to Vulnerability Assessment.* (London: Earthscan).

Krankina, O.N., Dixon, R.K., Kirilenko, A.P. and Kobak, K.I. (1997), 'Global Climate Change Adaptation: Examples from Russian Boreal Forests', *Climatic Change* 36: 197–216.

Kulyasova, A. and Pchelkina, S. (2004), 'Global Processes in Local Scale: Malashuika Case Study', *Spectrum* 3. (in Russian).

Kulyasova, A. and Kulyasov, I. (2005), 'Russian National Voluntary Forest Certification (Vozhegales Case Study)', *Spectrum* 1. (in Russian).

Kulyasova, A., Kulyasov, I. and Pchelkina, S. (2005), 'Regional Aspects of the Global Forest Certification Process', *Region: Economy and Sociology* 4. (in Russian)

Maracchi, G., Sirotenko, O. and Bindi, M. (2005), 'Impacts of Present and Future Climate Variability on Agriculture and Forestry in the Temperate Regions: Europe', *Climatic Change* 70, 117–135.

Moench, M. and Dixit, A. (eds) (2004), *Adaptive Capacity and Livelihood Resilience. Adaptive Strategies for Responding to Floods and Draught in South Asia.* The Institute for Social and Environmental Transition, International, Boulder, Colorado, USA and the Institute for Social and Environmental Transition, Nepal.

O'Brien, K.L., and Leichenko, R.M. (2000), 'Double Exposure: Assessing the Impacts of Climate Change within the Context of Economic Globalization', *Global Environmental Change* 10, 221–232.

Saelthun, N.R. (1995), 'Climate Change Impact on Hydrology and Cryology', pp. 45–49 in Guisan, A., Holten, J.I., Spichiger, R. and Tessier, L. (eds).

Smit, B., Burton, I., Klein, R.J.T. and Wandel, J. (2000), 'An Anatomy of Adaptation to Climate Change and Variability', *Climatic Change* 45:223–251.

Smit, B. and Pilifosova, O. (2001), 'Adaptation to Climate Change in the Context of Sustainable Development and Equity', pp. 877–912 in IPCC [J.J. McCarthy, O.F. Canziani, N.A. Leary, D.J. Dokken and K.S. White, eds]

Smit, B. and Skinner, M.V. (2002), 'Adaptation Options in Agriculture to Climate Change: A Typology', *Mitigation and Adaptation Strategies for Global Change* 7, 85–114.

Smit, B. and Wandel, J. (2006), 'Adaptation, Adaptive Capacity and Vulnerability', *Global Environmental Change* 16, 282–292.

Watson, R.T., Zinyowera, M.C., Moss, R.H. (eds) (1998), *The Regional Impacts of Climate Change: An Assessment of Vulnerability* (Cambridge: Cambridge University Press).

Chapter 12

Regional Governance, Path Dependency and Capacity-Building in International Environmental Cooperation

Monica Tennberg

Introduction

Since the early 1990s a complex structure of regional environmental governance has been developed for Northwest Russia that has included hundreds of projects to support Russian environmental capacities. International environmental cooperation in the region is a product of many – in most cases separate – international political initiatives by different political actors in the region that reflect their various environmental, economic and political interests in Northwest Russia and the Russian Federation in general. However, at the same time, Russian local, regional and national capacities to tackle environmental issues may well be lost due to economic constraints, continuous administrative reforms and a lack of commitment to sustainable development. These Russian domestic developments undermine international efforts and account in part for the modest performance of international cooperation in the region. In this chapter, the two parallel developments – international and domestic – are studied from the perspective of regional environmental governance, path dependency and environmental capacity-building. The emerging path in international environmental cooperation establishes the region as an object of action, not as an environmental actor itself. The domestic developments in Russian environmental policy have also led to diminished capacities for environmental action in the region. Path dependency in Russian environmental management constrains the international environmental cooperation in the region.

State Still Needed

Governance is, according to the Commission on Global Governance (1995, 2–3), 'the sum of the many ways individuals and institutions, public and private, manage their common affairs'. In the case of Northwest Russia, the governance structure is regional and includes the establishment of international political bodies and funding programmes that bring together various actors from the local, regional and national levels to tackle environmental concerns. The idea of governance for the environment emphasizes the importance of action at various levels and especially the importance of non-governmental actors in addressing environmental issues. Under the conditions of globalization and global governance, national environmental authority is no longer the sole agent responsible for designing, implementing and solving environmental problems (Sonnenfeld and Mol 2002, 1325). This chapter, however, argues that the state is an important environmental actor in the development, regulation and implementation of international environmental cooperation. Highlighting the importance of the state as an environmental actor and its environmental capacities both domestically and internationally challenges the often somewhat restricted approach in the study of global environmental governance that focuses on inter-governmental developments only (Speth and Haas 2006). Looking into the state's environmental capacities broadens the object of study for scholars of international relations in examining the performance of international environmental cooperation (see Schreurs and Economy 1997; DeSombre 2000; Young 2002).

The aim of regional environmental cooperation in general is to improve Russian environmental capacities to tackle the country's environmental problems, which are considered to be self-inflicted and in many cases a result of years of particular development policies. According to the 1999 OECD assessment of Russian environmental performance, 'Russia's environmental problems are rooted, in large measure, in the problems prevalent in its society today: weak rule of law, corruption, the lack of a civil society, large-scale economic problems while energy and water subsidies encourage wastefulness'. However, these environmental problems have been perceived as threatening Russia's neighbour countries as well, especially in the late 1980s. Environmental capacity-building refers to the development of human, organizational, technical, material and institutional resources to deal with environmental concerns. At a more detailed level, environmental capacity-building includes the strength, competence, and configuration of governmental and non-governmental actors in developing and implementing environmental policies. It also embraces various framework conditions for environmental action, including cognitive-informational, political-institutional and economic-technological conditions. The utilization of environmental capacities depends on the strategy, will and skill of actors and their opportunities for action in particular situations. These capacities have to be related to the kind of problem at hand, its urgency, complexity, and the power resources related to it, such as possible alliances and supporters of the particular issue and its solutions (Weidner and Jänicke 2002, 4–6).

In general, much more attention has been focused on national capacity-building than international capacity-building for environmental issues. However, the issue is important in international environmental cooperation. Without attention to capacity-building, international environmental cooperation may become 'mere sets of good intentions without the ability to meet international commitments and to comply with international norms and principles' (Vandeveer and Dabelko 2001, 18). National capacity-building is described as a 'necessary condition for the effectiveness' of international environmental institutions (Levy et al. 1993, 415). In the study of international environmental cooperation, most of the attention has been focused on the capacity of an international environmental regime to fulfil the aims set for it. In the study of institutional arrangements the approach to international capacity-building seems quite formal. It concentrates on questions such as 'what laws are passed, what new policies are adopted, what the extent of their enforcement is, and what funds are invested' (Keohane et al. 1993, 16).

Environmental capacities play an important role in the everyday management of international environmental projects, but the development of national capacity in general also creates a context for the success or failure of international cooperation in terms of the availability of domestic funding arrangements, administrative and legal rules for international cooperation, and national commitment to the goals of the cooperation. In his study on East-West environmental cooperation, Robert G. Darst (2001, 206–207) suggested that economic, political and administrative capacities play an important role in international environmental cooperation. The recipients of international environmental assistance, for example, have to be able to participate in the planning and financing of joint projects and to contribute both legally and administratively to implementing them. In the case of Northwest Russia, Darst concluded, the recipients' capacity to produce such resources depended on the overall political and economic situation and on the extent of control the participants had in international environmental cooperation. The diffusion of responsibility between different actors regionally does not mean that these actors are automatically capable of successfully planning and implementing international environmental projects.

Capacity-building is not 'necessarily a steady, linear process' (Weidner 2002, 1342). According to the ecological modernization theorists, sufficient societal, political, administrative and organizational capacity is needed for 'ecological modernization' in Eastern Europe and in Russia. International environmental assistance as such is not enough for ecological modernization. Andersen (2002) suggests two different hypotheses on the basis of this idea: 1) a more optimistic expectation that ecological modernization will take place in accordance with new political and economic realities and with foreign assistance and 2) a more cautious evaluation of the importance of domestic political processes which either support or hinder political and economic reforms needed for ecological modernization. In his analysis of various Eastern European countries and their environmental capacity-building, the situation in Russia is much more difficult compared to others in political terms, as a viable framework for environmental financing and policy making has generally not been developed. Both in terms of indicators of ecological modernization and environmental capacity-building, Andersen (2002) ranks Russia in the bottom group among the Eastern European countries. In his view, there is an intricate relationship between domestic capacity-building processes, foreign assistance and broader economic and political conditions.

The capacities and actions of various actors are controlled by institutions. These institutions enable, constrain and control the capabilities of actors. The concept of path dependency relies on the notion that 'small and chance circumstances can determine solutions that, once they prevail, lead to a particular path' (North 1990, 94). Institutions show path dependency because their development is ruled a particular set of opportunity structures, including formal and informal constraints. As North suggests, institutional change is often incremental and marginal. The concept of environmental capacity-building suggests that in some cases locked-in trends of developments can be changed and new development paths started. International environmental cooperation implemented in various projects in Northwest Russia – including technical assistance, training and institution-building – aim to support change in Russian environmental management practices. However, as Blokker (2005, 509) points out, one important point should be considered carefully: transition studies have been too occupied with assessing the convergence of post-communist countries with European standards. Path dependency theory points to the distinctiveness of former Eastern European countries and their experiences. Political reforms are seen as path-dependent processes based upon institutionalized forms of learning and struggles over pathways that emerge out of the intersection of old and new. Current and future developments cannot be understood unless the past is reflected upon, so that the constraints and possibilities for current transitions, become clear.

In this case, the most important paths to analyze are, on the one hand, the established practices of international environmental cooperation in Northwest Russia, which emphasize the importance of international funding, the use of Western expertise and high-level political cooperation. On the other hand, the history of environmental management at the domestic level in Russia is characterized by organizational reforms, inter-agency struggle, limited funding and the influence of various political interests (Peterson 1995; Peterson and Bielke 2001; Crotty 2003; Oldfield 2005). These two institutionalized practices create a particular form of international environmental cooperation that constrains environmental capacity-building in Northwest Russia.

This study[1] also suggests a similarly close relationship between domestic Russian developments and the international efforts to create environmental capacity-building. In the following, the capacity-building processes, especially those involving the development of organizational, material, and human resources, are described and discussed. The chapter is based on information available from various officials, governmental reports and interviews conducted among Nordic and Russian participants in international environmental cooperation – especially Arctic, Barents Euro-Arctic and EU Northern Dimension cooperation – in Northwest Russia in 2003–2004 and 2006. In addition to interviews, a survey was made among project managers working in Northwest Russia in 2004 and

---

1   I am grateful to my Russian colleagues, especially Larissa Ryabova, Lyudmila Ivanova, Nadezhda Kharlampieva, Tamara Semenova, Antonina Kulyasova and Svetlana Agarkova, who helped me in organizing the interviews and who translated them. The interview trips were funded by travel grants from the Academy of Finland.

2006. The interviews provide an insider view of the developments in international environmental cooperation in the region. The quotations from the interviews here are chosen to shed light on the experiences of the actors involved.

Organizational Resources for the Environment

The governance structure for international environmental cooperation in Northwest Russia is characterized by international environmental projects that are coordinated by multilateral political forums and funded by international funding agencies. Many multilateral forums have been working in the region since the early 1990s. In these forums, the environment is only one of areas of cooperation. The Council of the Baltic Sea States (CBSS) is a political forum for regional intergovernmental cooperation comprising the states of the region and the European Commission. The Barents Euro-Arctic Council (BEAC) is a forum for international cooperation that includes the Nordic countries, Russian authorities and a representative of the European Union. The Barents Regional Council also brings together representatives of the county governors and their counterparts. The Nordic Council of Ministers added Northwest Russia to its programme in the mid-1990s and included cooperation on environmental issues. The Arctic Council was established in 1996 as an intergovernmental forum to discuss common concerns. It has also developed as a forum of international cooperation between governments and indigenous peoples in the Arctic. Finally, environmental issues have had an important place in the Action Plans of the EU Northern Dimension (2001–2003, 2003–2006). The Northern Dimension Environmental Partnership (NDEP), established in 2001, supports cooperation in environmental protection and nuclear safety in the region.

Both the Nordic and Russian interviewees value the efforts in the last decade to build multilateral structures and networks for international cooperation. The interviewees stressed that the different multilateral platforms have their own functions and roles, although some of them thought that the roles overlap and that the various activities need to be rationalized. One Nordic interviewee pointed out that the work to strengthen these structures and their effective use should continue, and nothing should be done to weaken them. For example, better cooperation between decision-makers and funding agencies is needed in the region. International environmental cooperation in the Barents region was valued especially by the Russian interviewees. For them, Barents cooperation was the real thing, 'real cooperation,' and the Norwegians had the resources to make the cooperation work in practice. According to a Nordic interviewee, it provides a space for a 'good dialogue' with the Russian partners. In comparison, Arctic cooperation does not provide such a forum due to its structure: there are few meetings, with large delegations. In the case of the EU Northern Dimension, a Nordic interviewee noted that 'the Russians have not been consulted on it in a proper way' and therefore there are no high expectations for future success. The Northern Dimension is something that 'is implemented for the Russians and for the Russian Federation'. The criticism of the Northern Dimension can be partly understood in light of the Russian experiences of TACIS projects. The EU-funded TACIS projects received the greatest amount of criticism from the Russians. The interviewees considered many EU projects as 'just paper.' However, it must be noted that during the second round of interviews in 2006, Russian statements about the EU Northern Dimension were in general much more positive than in 2003–2004.

One particular feature in these forums is that the role of NGOs is marginal, especially in the case of Russian NGOs. International NGOs expanded their activities in Northwest Russia in the late 1980s, among them Greenpeace, Bellona and WWF. Some national NGOs, such as the Norwegian youth organization Nature and Youth have been working together with Russian environmentalists on the Kola Peninsula since 1989. The role of NGOs is more active in the implementation of various international projects in Northwest Russia. Lately, constraints have been imposed by legal means in Russia on the participation of non-governmental organizations in international cooperation. A new law which came into effect in April in 2006 set stricter registration and reporting requirements for NGOs.

The relationship between Nordic and Russian participants is quite strained at the higher political level. The Nordic interviewees described the Russian partners as 'not very committed' in international environmental cooperation. A Nordic interviewee defined the Russian participants' lack of commitment as the main problem in international environmental cooperation with them: 'A project needs to have an owner who wants to manage the project and its implementation.' This problem is encountered in the interaction with different Russian regions, in many projects, and in the interaction between individuals at different levels. At the higher political level, the norms of 'good partnership' were often violated, but in most cases the Nordic participants tolerated such incidents. One of the Nordic interviewees pointed out that 'we will continue working with the Russians, even if they do not really participate in cooperation'.

Some of these problems of commitment originate in administrative changes in Russian environmental management. The reforms of the Russian environmental administration have limited participation in regional environmental governance. In May 2000, President Vladimir Putin eliminated the State Environmental Committee altogether and placed its responsibilities and personnel under the Federation's Ministry of Natural Resources (Wernstedt 2002). In 2004, President Putin started a comprehensive reform to rationalize the government's administrative structures in the Russian Federation. In addition, the reform aimed at reducing the number of administrative units and staff in the federal bureaucracy. The reform has been followed by administrative re-organization at the regional level and a re-division of responsibilities between federal and regional authorities for the environment. These developments have made international environmental cooperation with the Russians difficult, as the personnel in charge and the rules of participation change constantly. This is also noted on the Russian side. One of the Russian interviewees described the situation as follows: 'Revolution never ends'. This creates problems for all parties: 'It is a permanent restructuring hell' which leads to uncertainty about who the responsible parties are and who to approach. To make things even worse, much of the expertise in various governmental bodies has been lost due to the experts' low salary and a lack of respect for their work. Many of them have transferred to work in various international organizations and programmes. However, many of the interviewees stressed that the most difficult years are now over.

Regional environmental governance aims at advancing sustainable development. At the moment, sustainable development seems to fit poorly into the domestic structure of norms and power in the Russian Federation. Many have suggested that the Russian idea of sustainable development has yet to be developed. Sustainable development is interpreted by the Russian interviewees as 'a need to separate national economic interests and the economic interests of big companies, a problem of incomplete and imperfect legislation, a problem of a low level of public participation, or a lack of knowledge of the state of the environment and priorities for solving the problems.' However, it seems that an 'economic' interpretation of the idea of sustainable development has gained a legitimate position among the Russian participants in the Arctic cooperation when Russia was chair of the Arctic Council in 2004–2006 (Report of senior arctic officials to ministers at the 5th Arctic council meeting 2006). In the case of the Barents Council, the norm of sustainable development was considered self-evident among the participants at the beginning, but it has become somewhat problematic after being challenged by the Russian participants. It is clear that they often prefer economic cooperation to environmental cooperation. One interviewee stated: 'First the Western interests, and then ours – our time will come'. In the case of the NDEP, Russian resistance to the norm has been vocal. According to the Russian interviewees, 'concern for the environment gets too much attention at the expense of other important issues, such as economic development and growth'.

Funding for International Environmental Cooperation

One of the most characteristic features of international environmental cooperation in Northwest Russia is international environmental assistance from various bilateral and multilateral sources in the form of grants, loans and investments. According to the OECD (2003, 39–40), the Nordic countries along with France, Germany and the UK account for 80 per cent of the international assistance to Russia. In 1990, the Nordic Council of Ministers established the Nordic Environment Finance Corporation (NEFCO), which provides grants to important multilateral projects in the nearby areas of Russia that reduce environmental damage and pollution. In December 2004, the governments of Finland, Iceland, Norway and Sweden agreed on a special ad-hoc financing facility, the Barents Hot Spot Facility, whose capital of approximately 3 million euros is for project development. The facility is managed by NEFCO. However, one characteristic of Nordic funding is that it is bilateral, not multilateral, in nature. Nordic multilateral funding has been only 6.2 per cent of the total sum used by the Nordic countries bilaterally (Pohjoismainen ministerineuvosto 2000, 36, 48). Since 1993 approximately 2,000 Norwegian-Russian projects in the Barents region have received support through the Barents Secretariat (Norwegian Ministry of Foreign Affairs 2003, 33). Finland has supported more than 1,500 environmental projects in the region over the period 1990–2004, committing resources amounting to some 143 million euros (Ympäristöministeriö 2005).

The EU was identified as the main partner in international environmental cooperation in the region. EU assistance for Russia is provided mainly through the EU's technical assistance programmes. EU funds allocated for the Russian environment sector in 1996–2003 totalled around 60 million euros. Environmental funding is provided through other programmes as well. All technical assistance to Russian partners is in the form of grants, not loans. Russia has also been one of the main beneficiaries of the TACIS Nuclear Safety Program (EU Commission 2001). In order to support investments in the region, the EU has allocated 50 million euros for the Nordic Dimension Environmental Fund, of which 40 million euros is earmarked for nuclear waste related projects and 10 million euros for environmental projects. The EU Commission is the primary contributor to the fund but Denmark, Finland, Germany, the Netherlands, Norway, Russia and Sweden have all made contributions to it as well.

The need for any type of funding during the difficult years in the Russian economy in the 1990s explains the Russian willingness to cooperate with Western partners in environmental matters. A low level of environmental expenditure has continued since the early 1990s. In the 1990s the Russian economy went through a process of stabilization and privatization with many problems of adaptation. The debts inherited from the Soviet Union have also strained the economy, with the outstanding loans limiting opportunities for domestic investments or additional loans (Kjeldsen 2000; Wernsted 2002; OECD 1999). Lately, despite its economic growth, the Russian Federation has politically prioritized payment of its debts, meaning that that very little domestic funding has been available for environmental protection measures. In Northwest Russia, all regions provide statistical information about their environmental expenditure and investments. However, according to a World Bank report (2004), federal expenditure for environmental protection is still low compared to other federal countries. The lack of financial commitment to environmental issues is explained by one of the Russian interviewees: 'It is the Russian priority to improve the economic situation. This is the first task of politicians. Russia is in a completely different position compared to other countries.'

In general, Nordic and Russian commitment to future environmental cooperation was strong on both sides, although its forms and the allocation of responsibilities between the participants will be changed. One of the Russian interviewees pointed out that 'the time for consulting is over'. There was a widespread view among the interviewees that due to the recent growth of the Russian economy there should be more Russian intellectual and material resources available to take care of domestic environmental concerns. Russian input in international environmental cooperation cannot consist solely of in-kind services, as has been the established practice so far. New ways of organizing international environmental cooperation and the funding for it between the partners are also needed. In 2006, a project manager reported positive developments in this respect: 'There has been a marked, significant change in the level of local Russian funding. The conscious attempts to equalize the relationship have worked out well. Russian partners understand that it is their needs that are important.'

The Importance of the 'Human Factor'

The political and economic organization of the regional environmental governance structure forms a particular institutional framework for international environmental cooperation in the region. It both constrains the actors and enables them to work for the environment. The results of the study highlight the importance of human resources in international environmental cooperation. According to the participants in international environmental cooperation in Northwest Russia, the conditions for successful projects include having 'the right people in the right places with good, stable, and trustful relationships'. According to both the Nordic and Russian interviewees, funding as such was not a critical factor in assessing the conditions for performance in international environmental cooperation, although a successful project requires proper funding arrangements. The success of a project depends on the project leaders' capacity to deal with administrative difficulties, cultural differences and different working practices. Failure can be expected when people only have money in mind. In addition, international environmental projects will be less successful if the participants perceive the arrangements for cooperation as unequal, when the cooperation lacks clear goals and behavioural criteria for all partners, when it has no general political support or if there is no common interest as the basis for the project.

The importance of the human factor is further emphasized in the results by the importance of sharing ideas, experience and knowledge. In terms of tackling environmental problems, the interviewees pointed out that 'the lack of information is the biggest problem'. The Russian interviewees seem to disagree on this point among themselves. One of the interviewees noted that 'we don't have enough specialists.' However, another stressed that 'the level of Russian scientists is not lower than that of Western consultants.' On the other hand, in many cases, the Russians feel that their expertise and experts are not acknowledged and respected. The Westerners' attitude towards their Russian partners was also criticized: 'The Westerners do not share information and listen to their Russian partners enough.' There is also a lot of disinformation around. One of the Russian interviewees pointed out: 'our mass media do not give us a real picture about the environmental situation in the country'. The Western media are criticized for producing a one-sided and biased image of Northwest Russia as an 'ugly, polluted and spoilt' region.

On the other hand, the Nordic interviewees complained that the Russian participants were not willing to discuss openly their environmental situation and problems. They prefer to stay quiet or talk about business affairs, such as economic cooperation, instead of environmental concerns. The Nordic interviewees stress that the aim of international environmental cooperation in the region is not really to tackle 'transboundary' environmental problems but to assist Russia to deal with its domestic environmental problems. 'In the end,' some of the Nordic interviewees pointed out, 'the Russians have to solve their problems themselves.' Some of the Russian interviewees stress that 'only the Russians can solve their own problems'. Some of the Western partners were also accused of being hypocritical – expressing concern for the state of the Russian environment 'just because the environment in Western countries has already been destroyed'.

In general, the development of regional environmental governance has produced a Western-led cooperation, dominated by Western interests, concerns and funding. This has been experienced as unequal and unfair by the Russian interviewees. In their opinion, despite the visits and studies conducted, Western consultants and their reports do not always 'result in positive outcomes' or produce any 'real kind of international environmental cooperation'. The minimal role of Russian expertise and experts is raised as an issue by many interviewees – especially the lack of opportunities for Russian experts to participate in international environmental cooperation and the low respect internationally for Russian experts and expertise. The use of Western funding legitimizes the use of Western experts and expertise to find solutions to local and regional problems.

The Emerging Path in Regional Environmental Governance

As a whole, the Nordic interviewees regard Northwest Russia as 'not an easy place for cooperation' due to various economic, political, administrative and cultural constraints in the region. Many of the Nordic interviewees expect that the period of international environmental cooperation with Northwest Russia has come to an end in its current form. In particular, the future of the EU-funded international environmental cooperation is open. Despite the lack of clear evidence for an improvement in the state of the environment due to international efforts in Northwest Russia, most of the Nordic interviewees were committed to continuing the cooperation. They expect that all these efforts will bear fruit when the political and economic situation in the Russian Federation becomes better. This will require both time and patience. Long-term benefits from the international environmental cooperation can be expected within ten to twenty years. On the Russian side, a widely shared expectation among the interviewees was that international environmental cooperation between the Western and Russian partners will continue into the future at least at the same level as before or develop even further.

The regional environmental governance developed for Northwest Russia since the early 1990s is characterized by multilateral forums of international cooperation, various funding mechanisms and international environmental projects. In general, the interviewees considered the environmental performance of this cooperation modest. The project managers have a somewhat more positive assessment of the results of international environmental cooperation in the region. A point which is worth considering in the light of the results of this study is the need for Western actors to consider their own approaches and strategies in international environmental cooperation. The use of experts and expertise from both sides seems to be a critical question that needs to be settled in the future. Weidner (2002, 1358) points out in his analysis that 'the transfer of Western instruments and innovations is not unproblematic'. In this particular case, there have been many constraints on the effective use and development of organizational, material and human resources and capacities for the environment, both formal and informal.

These results suggest that Russian environmental capacities have not been strengthened by international environmental cooperation. Partly this may be a consequence of practices in international environmental cooperation which place Northwest Russia into the position of an object, instead of a subject, of international environmental cooperation. Yet, this is also to large extent a consequence of the domestic Russian developments in environmental policies. Analyzing the regional environmental governance and its performance without taking into account domestic developments tells us only half the story. International efforts in environmental capacity-building in Northwest Russia have been stalled by Russian environmental politics, which is characterized by inadequate domestic environmental expenditure, constant administrative challenges, a lack of opportunities for Russian experts to participate in international environmental cooperation and a low level of commitment to sustainable development. This emerging path and changing it in the regional environmental governance is a challenge for the future of the cooperation.

References

Andersen, M.S. (2002), 'Ecological modernization or subversion? The Effect of Europeanization on Eastern Europe', *American Behavioral Scientist* 45:9, 1394–1416.

Blokker, P. (2005), 'Post-Communist Modernization, Transition Studies, and Diversity in Europe', *European Journal of Social Theory* 8:4, 503–525.

Commission on Global Governance (1995), *Our Global Neighbourhood* (Oxford: Oxford University Press).

Crotty, J. (2003), 'The Re-Organization of Russia's Environmental Bureaucracy. Regional Response to Federal Changes', *Eurasian Geography and Economics* 44:6, 462–475.

Darst, R.G. (2001), *Smokestack Diplomacy. Cooperation and Conflict in East-West Environmental Politics* (Cambridge, MA: MIT).

Desombre, E. (2000), *Domestic Sources of International Environmental Policy. Industry, Environmentalists, and U.S. Power* (Cambridge, MA: MIT).

EU Commission (2001), *Communication from the Commission. EU-Russia Environmental Cooperation*, COM(2001) 772 final. http://delrus.cec.eu.in/en/images/pText_pict/231/COM_2001_0772.doc.

Haas, P.M., Keohane, R.O. and Levy, M.A. (eds) (1993), *Institutions for the Earth: Sources of Effective International Environmental Protection* (Cambridge, MA: MIT Press).

Keohane, R.O., Haas, P.M. and Levy, M.A. (1993), 'The Effectiveness of International Environmental Institutions', pp. 3–26 in Peter Haas et al. (eds).

Kjeldsen, S. (2000), 'Financing of Environmental Protection in Russia: The Role of Charges', *Post-Soviet Geography and Economics* 41:1, 48–62.

Levy, M.A., Keohane, R.O. and Haas, P.M. (1993), 'Improving the Effectiveness of International Environmental Institutions', pp. 397–426 in Peter Haas et al. (eds).

North, D.C (1990), *Institutions, Institutional Change and Economic Theory* (Cambridge: Cambridge Polity Press).

Norwegian Ministry for Foreign Affairs (2003), *Opportunities and Challenges in the North*. Report no. 30 to the Storting (2004–2005). http://www.regjeringen.no/upload/kilde/ud/stm/20042005/0001/ddd/pdts/stm200420050001ud_dddpdts.pdf.

OECD (1999), *Russian Environmental Performance Review* (Paris: OECD).

OECD (2003), *Task Force for the Implementation of the Environmental Action Programme for Central and Eastern Europe* (EAP). http://www.oecd.org/dataoeed/37/18/26732337.PDF.

Oldfield, J. (2005), *Russian Nature: Exploring the Environmental Consequences of Societal Change* (Aldershot: Ashgate).

Peterson, D.J. (1995), 'Russia's environment and natural resources in light of economic regionalization', *Post-Soviet Geography* 36:5, 291–310.

Peterson, D.J. and Bielke, E.K. (2001), 'The Re-Organization of Russia's Environmental Bureaucracy: Implications and Prospect', *Post-Soviet Geography and Economics* 42:1, 65–76.

Pohjoismainen ministerineuvosto (2000), 'Lähemmäksi pohjolaa. Esitys tarkistetuksi lähialuestrategiaksi', *TemaNord* 2000:572. http://www.Norden.org/naromraaden/fi/naermare-norden.asp#yhtenveto.

Report of senior arctic officials to ministers at the fifth Arctic Council ministerial meeting (2006), Salekhard October 26, 2006. http://archive.arcticportal.org/287/01/SAO-REPORTTO_MINISTERS.pdf.

Schreurs, M. and Economy, E. (1997), *The Internationalization of Environmental Protection* (Cambridge: Cambridge University Press).

Sonnenfeld, D. and Mol, A.P.J. (2002), 'Globalization and the Transformation of Environmental Governance', *American Behavioural Scientist* 45:9, 1318–1339.

Speth, J.G. and Haas, P.M. (2006), *Global Environmental Governance* (Washington D.C: Island Press).

Vandeveer, S.D. and Dabelko, G.D. (2001), 'It's Capacity, Stupid: International Assistance and National Implementation', *Global Environmental Politics* 1:2, 18–29.

Weidner, H. (2002), 'Capacity-Building for Ecological Modernization', *American Behavioral Scientist* 45:9, 1340–1368.

Weidner, H. and Jänicke, M. (2002) (ed.), *Capacity Building in National Environmental Policies* (Heidelberg, New York: Springer).

Wernstedt, K. (2002), 'Environmental Protection in the Russian Federation: Lessons and Opportunities', *Journal of Environmental Planning and Management* 45:4, 493–516.

World Bank (2004), *Environmental Management in Russia: Status, Directions and Policy Needs*. http://www.oecd.org/dataoecd/10/19/35486874.pdf.

Ympäristöministeriö (2005), *Ympäristöyhteistyön Venäjä-strategia 2006–2010*. http://www.ymparisto.fi/download.asp?contentid=42627&lan=fi.

Young, O.R. (2002), *Institutional Dimensions of Global Environmental Change* (Cambridge, MA: MIT Press).

Chapter 13
Summary

Soili Nystén-Haarala

The transition to a market economy – or at least the further development of such an economy – is still underway in Russia. Even if the present situation could be described as a 'Russian market economy' and 'Russian democracy', the change still seems to be a rapid one. Changes can be seen in official as well as unofficial institutions and in the interplay between them. In this volume we have applied the concept of governance, defined as the development from government to governance and, in the case of Russia more specifically, from uniform centralized decision making towards networking and interaction between different actors and levels of the government. In Russia, networking and self-governance have traditionally been means to survive when official institutions have failed. Accordingly, it is no wonder that transition and the resulting collapse of official institutions encouraged recourse to networks of unofficial institutions. Then again, decentralization of state power, its devolution to the regions and the introduction of municipal self-governance were on the agenda of transition of official institutions; but it seems that changes in official institutions were made too quickly and on too wide a front, and therefore produced inefficient results. Many of the changes in official institutions were not in line with unofficial institutions and therefore not internalized quickly enough.

The few steps taken back to more familiar circumstances – centralization, one-man rule, one-party politics and other path-dependent ideas of the superpower past that, it was assumed, would keep the country united – will hinder the development of modernity in Russia in the long run. As long as the Russian government does not adhere to past, path-dependent ideas for too long, the steps back can be interpreted as necessary ones taken in a confusingly quick transition. It should be borne in mind that Russia is still paying for the badly planned (or unplanned) privatization in the form of an oligopolistic and 'oligarch' driven economy, delayed restructuring of industry, corruption and a still prevailing 'grab and run' business mentality. In this perspective, resorting to past, path-dependent ideas becomes understandable.

It is not difficult to find failures in the Russian transition, but it is no wonder because the change has been so rapid and has in fact only been going on only for some two decades. The task of creating and developing new official institutions has been a huge and demanding one. For example, forest legislation, analyzed in this volume, has far-reaching objectives, such as the effectivization of forestry, on the one hand, and environmental challenges, such as sustainability, on the other. The government has had to deal with property rights issues, modernization of administration as well as demands by various business and civil society actors. The result appears to be policy in a whirl that looks like no policy at all.

Misuse of official institutions such as company or bankruptcy legislation has been a serious drawback, one that was difficult to anticipate or even imagine. Opportunistic behaviour was so successful in the Russian economy during the transition period that it is now difficult to get rid of. Such behaviour also reflects a deep disrespect of official institutions.

Then main challenge for the government is to try to create and maintain trust in the economy and the state administration as well as in all official institutions generally. During V. Putin's presidency, the government seems to have sought social trust by returning to old ideas and facades, which will not work in the long run. If the aim is to create a credible investment climate and economic competitiveness, no government can work such miracles without cooperation and the necessary social capital. Although moving from government to governance has its challenges in terms of accountability and legitimacy, it is the path that even the Russian government has to take to succeed in the long run. The era of nation-state empires has passed.

While the central government has struggled to take back its lost power, local actors have had to organize the running of local communities and businesses. With a lack of interest on the part of the federal state and the weakness (or non-existence) of municipal self-governance, companies have taken the lead. In this task they have required the support of the regional state power, municipalities, NGOs and trade unions. Being the only source of wealth, business has had a heavy burden and strong incentive to find well-functioning local governance systems. Typical of these new governance systems has been the Soviet heritage, on the one hand, and the assumed expectations of Western consumers, on the other. The impact of the global markets is brought to local circumstances with the business challenges of exporting companies. The most influential interest groups on the local level have been international ENGOs and to some extent the regional trade unions, which are gradually gaining more power. Local inhabitants seem to gain from this development as a side effect in towns and villages that have exporting companies with business potential. In remote fishing villages the circumstances are more difficult. With declining catches and a declining fishery, the local people seem to feel that the government is ignoring them. However, even these people try to affect the governance of fisheries in a way that could benefit the local people.

In remote declining areas – be they Norwegian, Finnish or Russian fishing villages – the inhabitants seem to have similar feelings of disappointment in their governments. In the EU countries they can blame the government for being more interested in fulfilling EU directives than taking care of local interests; in Russia they blame the government in Moscow for abandoning them. Global trends seem to impact remote areas more harshly, a reality for which the governments seem to be unprepared. Local circumstances should, however, be given more attention. Encouraging people at grassroots level to find new solutions will not work by returning to the "normal" circumstances of the tight state government of the Soviet past. Finding, creating and maintaining social capital is the key to governance that can create trust and lead to new prosperity or at least satisfactory solutions. The lead that the companies have now taken to maintain wealth in their communities in the absence of the welfare state is often a typical transitional strategy, but one that cannot continue for long without impinging on their efficiency and international competitiveness. However, cooperation between companies, NGOs and other interest groups without the influence of the state may also result in new solutions that could never have been developed with the lead of the government. Modernization of the state from its demanding a leading role to working as governance partner in networks of actors will be the key to development towards new dimensions of modernity, a competitive economy and a socially and environmentally sustainable society.

Index

adaptation 45, 171, 227, 229–231, 233, 240–242, 252
adaptive capacity 227–230, 240–241
administrative resources 114, 117, 119
antimodern society 15, 150

bankruptcy 106, 109, 113, 117–120, 123
banks 37, 39–41, 43, 45–48
berries 66, 190, 229, 235

capacity-building 245–248, 255
central bank 37–38, 40–43, 46–49
centralization 16, 19, 26, 80, 82, 100, 259
civil society 24, 60–61, 63, 71–73, 169–170, 176, 178, 246, 260
clear-cutting 65, 143, 178
climate change 5, 55, 227–230, 236, 239–241
    fishing 58, 77–78, 85–86, 88–90, 92–93, 95–96, 98–99, 197–221, 229, 231–233, 235, 236 237, 238, 241
    forestry 25, 31–32, 38, 44–45, 50, 56, 58–63, 68, 70–73, 109, 132–144, 163, 175, 177, 182, 184–185, 187–188, 190–191, 199–200, 202, 215, 221, 238, 241
    invasive species 239–241
    seasonality 227, 236, 241–242
    subsistence 217, 229–232, 235–236, 239, 241
    temperature change 236–237
coastal fisheries 88, 90, 94–95, 100, 238
complexity 34–36
consumer campaign 172–173, 175, 187, 191, 193
corporate standards 139
court proceedings 105, 118, 122

decentralization 69, 71, 79–80, 86, 99–100
(macro)economic policy 31, 33–34, 36–38, 40–42, 46–47, 49–50

economic system 34, 38
environmental requirements 63, 144

federal state 16, 21, 25, 79–80, 91
federalism 12, 17, 77, 79–81
firewood 141, 143, 205, 229, 234, 237
fiscal policy 41, 49
fishery policy 85, 88, 97
fishing communities 241
forest certification 26, 56, 60, 71, 132–134, 141–146, 163, 170, 175, 194, 220
forest code 4, 18, 25, 56, 62–72, 176, 234
Forest Experts Union 134
forest management 55–61, 64, 66–68, 70–72, 131–134, 140, 144–146, 174–176, 181, 187, 189, 194–195
forest policy 56, 69–71, 73, 188
Forest Stewardship Council (FSC) 133, 146, 175, 234
formal rules 15, 113

governance 10–13, 15, 19–21, 24–26, 33, 37, 39–40, 50, 56, 69–71, 73, 77–101, 105, 111, 120, 131–133, 140, 143–146, 151–155, 162, 169, 170, 172, 174, 228–229, 240, 245–246, 249, 251, 253–255
    corporate governance 11, 153–155, 162
    economic governance 10
    fishery governance 24, 85–89, 91–101
    forest governance 56, 69–71, 73, 131–133, 145–146
    good governance 11–13, 50
    hybrid governance 145
    local governance 101
    multilevel governance 12, 77–84, 91, 95, 98–99, 101
    natural resource governance 16, 37
    by networks 78
    private governance 10–11, 26, 98
    self-governance 12–13, 20–21, 26, 82, 151–152, 155

state governance 10–12, 25, 71, 91,
98–99, 143–145
without government 78

holding company (-ies) 23–24, 105, 107–
108, 110–111, 115, 133, 136–137
hybridity 132, 139, 144

imitation (democracy) 15, 161
industrial fisheries 77
inflation 34, 36–38, 41–50, 139–142, 159,
194, 231–234
informal rules 10, 15
institutional economics 10, 99, 150
institutions 9–10, 13–16, 20, 25–26, 33,
35, 39–43, 71, 78, 81, 84, 91, 101,
114, 124, 150–151, 170, 175–176,
197–199, 209, 211, 220–222, 228,
236, 245, 247–248
official (formal) institutions 10, 15,
25–26, 71, 150–151
unofficial (informal) institutions 14–15,
26, 71, 150
integration principle 56–57, 60–62, 67, 69
international cooperation 245, 247,
249–250, 254

joint-stock company 106–108, 113–114,
120–124

Kalevala National Park 170, 174, 178–181,
183, 187, 189–191
Kolkhoz (-y) 78, 86, 93–94, 96, 198–200,
202–222, 229–233, 235–236

law-making 56, 63, 70
Lespromkhoz (-y) 133, 135–136, 139, 182,
231

managed democracy 15
market campaign 169–176, 179, 181, 183,
193–195
modernity 12
money 34, 36–45, 47–50
municipal (local) self-governance (self-
government) 12–13, 20–21, 26,
82–83, 91–96, 99–101, 151–152,
155, 158, 208

municipal services 21, 152, 154, 161
mushrooms 24, 66, 235, 190, 229, 240

national system of voluntary forest
certification in Russia 133, 142
nation-state 12–13, 78, 81, 163
nature management 197–203, 209, 220–221
NGO space of place 10, 12, 24–26, 56,
59–61, 63, 70–71, 78–79, 84,
97, 99–100, 143–146, 169–180,
182–195, 200, 222, 250

official (formal) rules 14, 153, 155
old growth forest 60, 71, 141, 170–171,
174–175, 178–184, 186–188,
191–194, 234
ordered bankruptcy 117

path breaking 11, 14
path dependency 3–4, 10–16, 26, 171–174,
194–195, 197–200, 208, 218,
220–222, 245, 248, 259
path shaping 198
Pomors 197–203, 209–210, 212–213,
215–216, 218–222
post-modern society 15
privatization 9, 20–23, 62, 71, 107,
115–116, 120, 152, 158, 161, 252
(fishing) quota 85, 88–90, 92, 94–96,
98–99

risk 35, 38, 44–45, 50
Russian environmental management 248,
250, 254
Russian forest industrialists 24, 133–134,
145, 154

scales 131, 146, 152
selective felling 143
shareholder 106–107, 110–113, 115, 119,
120–124
small-scale coastal fishing 235
social responsibility 26, 141, 152, 161,
163–164, 174, 232, 234
stakeholder 11, 40, 59–61, 68–69, 71–73,
144, 159, 162, 169–170, 172–178,
183–184, 187, 189–192, 194–195,
229–230

sustainable development 4, 33, 50, 55–61, 66–67, 71–73, 97, 245, 251, 255

takeover 106, 110–124
theory of social systems 37, 50
transaction costs 10, 12, 14, 16, 171–172, 174, 194–195
transformation 9, 10, 24, 36, 71, 78–79, 85–86, 93–94, 96, 114, 131–136, 139, 142, 149, 153, 197–200, 202–203, 207, 220–221
transition 10–11, 14–15, 26, 39, 48, 56, 69, 72–73, 113, 140, 152, 163, 170, 183, 202–204, 248

transnational space of place 173–176, 179, 180, 183–184, 187, 192–195
transplant, 106, 113–114, 124
trust 11, 26, 33–37, 39, 42, 45, 49–50, 111, 149–151, 153–154, 157, 159–160, 162–164, 210, 213, 253

unofficial (informal) rules 14

vertical power 17, 18, 80–81
virtual economy 39, 152
vulnerability 227–231, 240–241

wood-working enterprises 107–108

Printed and bound by CPI Group (UK) Ltd, Croydon, CR0 4YY

21/10/2024

01777088-0016